PALGRAVE STUDIES IN THE HISTORY OF
SCIENCE AND TECHNOLOGY

James Rodger Fleming (Colby College) and Roger D. Launius (National Air and Space Museum), Series Editors

This series presents original, high-quality, and accessible works at the cutting edge of scholarship within the history of science and technology. Books in the series aim to disseminate new knowledge and new perspectives about the history of science and technology, enhance and extend education, foster public understanding, and enrich cultural life. Collectively, these books will break down conventional lines of demarcation by incorporating historical perspectives into issues of current and ongoing concern, offering international and global perspectives on a variety of issues, and bridging the gap between historians and practicing scientists. In this way they advance scholarly conversation within and across traditional disciplines but also to help define new areas of intellectual endeavor.

Published by Palgrave Macmillan:

Continental Defense in the Eisenhower Era: Nuclear Antiaircraft Arms and the Cold War
By Christopher J. Bright

Confronting the Climate: British Airs and the Making of Environmental Medicine
By Vladimir Janković

Globalizing Polar Science: Reconsidering the International Polar and Geophysical Years
Edited by Roger D. Launius, James Rodger Fleming, and David H. DeVorkin

Eugenics and the Nature-Nurture Debate in the Twentieth Century
By Aaron Gillette

John F. Kennedy and the Race to the Moon
By John M. Logsdon

A Vision of Modern Science: John Tyndall and the Role of the Scientist in Victorian Culture
By Ursula DeYoung

Searching for Sasquatch: Crackpots, Eggheads, and Cryptozoology
By Brian Regal

Inventing the American Astronaut
By Matthew H. Hersch

The Nuclear Age in Popular Media: A Transnational History, 1945–1965
Edited by Dick van Lente

Exploring the Solar System: The History and Science of Planetary Exploration
Edited by Roger D. Launius

The Sociable Sciences: Darwin and His Contemporaries in Chile
By Patience A. Schell

The First Atomic Age: Scientists, Radiations, and the American Public, 1895–1945
By Matthew Lavine

NASA in the World: Fifty Years of International Collaboration in Space
By John Krige, Angelina Long Callahan, and Ashok Maharaj

Empire and Science in the Making: Dutch Colonial Scholarship in Comparative Global Perspective, 1760–1830
Edited by Peter Boomgaard

Anglo-American Connections in Japanese Chemistry: The Lab as Contact Zone
By Yoshiyuki Kikuchi

Eismitte in the Scientific Imagination: Knowledge and Politics at the Center of Greenland
By Janet Martin-Nielsen

Climate, Science, and Colonization: Histories from Australia and New Zealand
Edited by James Beattie, Emily O'Gorman, and Matthew Henry

The Surveillance Imperative: Geosciences during the Cold War and Beyond
Edited by Simone Turchetti and Peder Roberts

Post-Industrial Landscape Scars
By Anna Storm

Voices of the Soviet Space Program: Cosmonauts, Soldiers, and Engineers Who Took the USSR into Space
By Slava Gerovitch

After Apollo? Richard Nixon and the American Space Program
By John M. Logsdon

Frontiers for the American Century: Outer Space, Antarctica, and Cold War Nationalism
By James Spiller

Improvising Planned Development on the Gezira Plain, Sudan, 1900-1980
By Maurits W. Ertsen

Post-Industrial Landscape Scars

Anna Storm

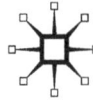

POST-INDUSTRIAL LANDSCAPE SCARS
Copyright © Anna Storm 2014

All rights reserved. No reproduction, copy or transmission of this publication may be made without written permission. No portion of this publication may be reproduced, copied or transmitted save with written permission. In accordance with the provisions of the Copyright, Designs and Patents Act 1988, or under the terms of any licence permitting limited copying issued by the Copyright Licensing Agency, Saffron House, 6-10 Kirby Street, London EC1N 8TS.

Any person who does any unauthorized act in relation to this publication may be liable to criminal prosecution and civil claims for damages.

Paperback edition published 2016. First published in hardcover 2014 by PALGRAVE MACMILLAN.

The author has asserted her right to be identified as the author of this work in accordance with the Copyright, Designs and Patents Act 1988.

Palgrave Macmillan in the UK is an imprint of Macmillan Publishers Limited, registered in England, company number 785998, of Houndmills, Basingstoke, Hampshire, RG21 6XS.

Palgrave Macmillan in the US is a division of Nature America, Inc., One New York Plaza, Suite 4500, New York, NY 10004-1562.

Palgrave Macmillan is the global academic imprint of the above companies and has companies and representatives throughout the world.

Hardback ISBN: 978–1–137–02598–2
Paperback ISBN: 978–1–137–58155–6
E-PUB ISBN: 978–1–137–02600–2
E-PDF ISBN: 978–1–137–02599–9
DOI: 10.1057/9781137025999

Distribution in the UK, Europe and the rest of the world is by Palgrave Macmillan®, a division of Macmillan Publishers Limited, registered in England, company number 785998, of Houndmills, Basingstoke, Hampshire RG21 6XS.

Library of Congress Cataloging-in-Publication Data

Storm, Anna.
    Post-industrial landscape scars / Anna Storm.
      pages cm
    Includes bibliographical references and index.
    ISBN 978–1–137–02598–2 (hardback : alk. paper)
     1. Industrial archaeology. 2. Industries—Environmental aspects.
  I. Title.
T37.S76 2014
363.73′1—dc23   2014016656

A catalogue record for the book is available from the British Library.

To my parents, Hans and Anne-Marie Nilsson

Contents

List of Illustrations   xi

Acknowledgments   xv

Chapter 1
Introduction   1
   Temporality and Significance   3
   Scars, Wounds, and Related Metaphors   4
   Heritage and Scars   5
   Post-Industrial Landscapes   8
   Industrial Heritage   10
   Localized Utopian Visions and Conflict Zones
      (the Places of the Book)   12
   Categories of Post-Industrial Landscape Scars   16

Chapter 2
Unstable Mountain   21
   Rich Mountains of the North   23
   An "Expendable Community"   26
   The Town and the Company at the Turn of the
      Twenty-First Century   31
   The Company Land and Local Heritage Processes   35
   Kiruna on the Move   37
   The Pit—Simultaneously "Feeding" and "Eating"
      the Town   40
   Concluding Remarks   45

Chapter 3
Distance of Fear   47
   A Christmas Gift   48
   Shaping Local and Regional Trust   52
   Antinuclear Winds   56

    A Danish View over the Sound   58
    Chernobyl and a "Premature" Shutdown   61
    The Spirit of Barsebäck   64
    Closing Down   67
    Significance and Future Use   69
    Concluding Remarks   73

Chapter 4
Lost Utopia   75
    The Plant and the Town   76
    Soviet Nuclear Activities   80
    Chernobyl   82
    The Žemyna Club and Sąjūdis   83
    Ignalina and Sniečkus within New Borders   86
    Improved Nuclear Safety and a Sacrifice to the West   88
    Nuclear Fear and a Long-Lived Heritage of
        Radioactive Waste   92
    Memory Work   94
    Concluding Remarks   98

Chapter 5
Industrial Nature   101
    Industrial Nature and the Ruhr District   103
    Abandonment, Discovery, and Spectacle   106
    Heritage and Nature in Coexistence or in Conflict?   112
    An Alternative Beauty of the Post-Industrial Landscape?   117
    A Possible Substitution Story   122
    Concluding Remarks   125

Chapter 6
Enduring Spirit   127
    A Place in the Forest, along the River   128
    Early Heritage Recognition   131
    The European Bison   132
    The Spirit of the Company Town   133
    A New Owner of the Old Industrial Area   134
    A "Cultural Team of Workmen" and a Municipal Project   136
    The New "Acropolis"   141
    Shaping Local Views   145
    The European Bison, Part Two   148
    Concluding Remarks   150

Chapter 7
Prospective Scars Unfolding   153

Notes	159
Bibliography	203
Index	223

Illustrations

Maps

1.1 The post-industrial landscape scars of this book are located in the Baltic Sea Region, in the countries of Lithuania, Germany, Denmark, and Sweden. 13
2.1 Present-day Malmberget and its surroundings. 22
3.1 Barsebäck nuclear power plant is located on a cape in Öresund, close to the cities of Malmö, Lund, and Landskrona. 49
4.1 The Ignalina nuclear power plant was built in the Soviet republic of Lithuania, close to the borders of Latvia and Belarus. 76
5.1 The Ruhr district is heavily industrialized and densely populated. 102
6.1 The old industrial area of Avesta was located along the source of its energy, the Dalälven river. 130

Figures

1.1 A former Soviet military shipyard in Karosta, Latvia, raises questions about imperial dreams and hard work, lost jobs and vanished communities, as well as about contamination, new economies, and aestheticizing industrial romanticism. 2
1.2 A team of workmen is on its way back from a shift Chusovoi, Russia. 7
2.1 Drilling in Malmberget around 1965. 26
2.2 Malmberget's present town center was built in the 1960s. 32
2.3 "The Pit" is a crucial feature of Malmberget. 40

2.4	As an experiment, a number of modern villas were moved from Malmberget to safer ground in Gällivare.	43
3.1	As a token of the close and friendly relations between Denmark and Sweden in the energy business, a Danish girl orchestra plays at the 1977 topping out party for reactor number two at Barsebäck.	53
3.2	Initially, the Barsebäck nuclear power plant was planned to consist of four, or even six, reactors.	54
3.3	Danish banners were common in the yearly antinuclear marches from Lund to Barsebäck in the late 1970s and early 1980s.	60
3.4	At the Danish National Museum in Copenhagen, visitors can learn about the history of the antinuclear movement.	72
4.1	The mono-industrial town of Sniečkus has in retrospect been labeled an "elite version" of Soviet planned space.	78
4.2	To work at the Ignalina nuclear power plant was a well-paid, high-status occupation for highly educated physicists and engineers.	79
4.3	In spite of protests from Lithuanian scientists, the construction of a third reactor at the Ignalina nuclear power plant began in 1985.	84
4.4	In the town center of Visaginas is a pedestal with the town symbol, a crane, on top.	93
5.1	When the former ironworks in Duisburg opened as a landscape park, visitors could climb one of the blast furnaces and get an extensive view, both of the post-industrial site and of the surrounding landscape.	107
5.2	Among many new activities in the landscape park, the local division of the German Alpine Club has constructed climbing paths in the former ore bunkers.	109
5.3	Traces of young lovers at the top of a blast furnace in the former ironworks, indicating its transformation into a place for adventure and refuge.	111
5.4	Spontaneous overgrowing as well as regular planting have become conscious strategies at many former industrial sites in the Ruhr district.	120
6.1	Avesta ironworks in 1912 with the Dalälven river to the right.	129

6.2 The old blast furnace plant was built in the 1870s out of slag stone brick, a by-product from the ore melting process. 137
6.3 As one part of the transformation of the old industrial area in Avesta, a sheet rolling mill was torn down. 143
6.4 The company symbol, a bison sculpture in stainless steel, initially had its place just outside the ironworks. 149

Acknowledgments

It was during a bus trip in Odense, Denmark, on the way to a conference venue that Jim Fleming suggested I should make a book out of my idea of post-industrial landscape scars. Thank you, Jim, for that crucial encouragement! Many other individuals, friends, and colleagues thereafter—and before—have decisively contributed to the making of this book.

First of all, I am deeply grateful to all of you who shared your time and stories with me in interviews: Lennart Johansson, Tommy Nyström, Lennart Thelin, Jørgen Steen Nielsen, Maria Taranger, Åsa Carlsson, Lennart Daléus, Håkan Lorentz, Bo Malmsten, Roland Palmqvist, Leif Öst, Zigmas Vaišvila, Victor Ševaldin, Algimantas Degutis, Alfredas Jomantas, Gunnar Johansson, Jan Nistad, Saulius Urbonavičius, Wolfgang Ebert, Ulf Berg, Jan Burell, Lars Åke Everbrand, Anders Hansson, Kenneth Linder, Karin Perers, Per-Erik Pettersson, Jan Thamsten, and Åke Johansson.

For supporting and facilitating my work by sharing personal contacts or by providing, interpreting, and helping out with empirical material, I warmly thank Marija Drėmaitė, Paul Josephson, Tatiana Kasperski, Åsa Dahlin, Ann-Marie Westerlind, Örjan Hamrin, Kersti Morger, Axel Ingmar, Tarmo Pikner, Anneli Sjölander-Lindkvist, Lars-Erik Jönsson, Johan Aspfors, Michael Farrenkopf, Ingo Kowarik, Michael Kleyer, and Maths Isacson.

For invaluable assistance in navigating archives and museum collections, I am indebted to Cecilia Hägglund, Anders Cederberg, Maria Taranger, Marija Drėmaitė, Magdalena Tafvelin Heldner, Lars K. Christensen, and Krister Källén.

For lending me their cars and sometimes accompanying me to fieldwork sites near and far, and for offering the very best accommodation, I wholeheartedly thank Aida Štelbienė, Ginas Štelbys, Siw Houltz, Hans Nilsson, and Anne-Marie Nilsson.

ACKNOWLEDGMENTS

The illustrations in the book are the result of several people's efforts and kindness. My warm thanks to Stig Söderlind for drawing the maps, to Karolis Kučiauskas for finding and scanning pictures at the Energy and Technology Museum in Vilnius, to Johannes Nilsson for finding and photographing the mound of stones at Barsebäck, to Peter Schmitt for rephotographing at the Landschaftspark Duisburg-Nord, and to all photographers who gave me permission to use their great pictures, although not all of these ultimately made their way into the book.

For funding my research, I want to express my sincerest gratitude to the Swedish Research Council Formas, the Baltic Sea Foundation, Vattenfall, the Swedish Research Council, the Swedish National Heritage Board, Marcus and Amalia Wallenberg Foundation, and Marianne and Marcus Wallenberg Foundation.

The daily work has been carried out in friendly and inspiring environments at the Department of Human Geography at Stockholm University, at the Centre for Baltic and East European Studies at Södertörn University, and at the Division of History of Science, Technology and Environment at the Royal Institute of Technology. Together with the Institute of Contemporary History at Södertörn University, these environments and their seminars have provided arenas for stimulating chats, crucial discussions, and friendships!

The fieldwork in Malmberget was carried out together with Krister Olsson, Gabriella Olshammar, Ingrid Martins Holmberg, Beate Feldmann Eellend, and Birgitta Svensson. Thank you all for the inspiring and joyful cooperation!

For reading and commenting on different chapters I am deeply grateful to Arne Kaijser, Kristina Šliavaitė, Paul Josephson, Fredrik Krohn-Andersson, Tatiana Kasperski, Florence Fröhlig, Peter Schmitt, Norbert Götz, Jim Fleming, Jennie Sjöholm, Sverker Sörlin, Dietrich Soyez, Krister Olsson, Hans Nilsson, Lars-Erik Jönsson, Magnus Rodell, and an anonymous reviewer. Your remarks have been invaluable, and I hope you will discern some traces in the final text due to your generous readings.

For the development of the scar metaphor, significant ideas and aspects were suggested by Anders Houltz, Jim Fleming, Stina Bengtsson, Anna Åberg, Eva Silvén, Torbjörn Lundqvist, Nina Wormbs, Cecilia Åse, Anthony Ince, and Michael Gilek. Thank you! It has been a pleasure to scrutinize the possibilities of a metaphor with you around.

Chris Chappell, Mike Aperauch, Sarah Whalen, and Bhavana Nair have been my companions at Palgrave Macmillan, proving that fruitful cooperation is possible across the Atlantic divide. Per Högselius's valuable advice on the publication process and Susan Richter's impressive editing of the text at a very late hour were both critical and most appreciated.

Anders Houltz, best reader ever. How many versions of this book have you carefully commented on and decisively improved? Now it is time to celebrate. May I have a tango?

Chapter 1

Introduction

I am sure you have seen them—post-industrial landscape scars. It might be a mountain irrevocably turned into an open-pit mine, surrounded by slag heaps. It could be polluted ground, abandoned, overgrown, and perhaps forgotten. It may be a dilapidated factory in a fading mono-industrial town, or in the midst of an otherwise bustling urban area, presenting a dangerous but attractive playground (figure 1.1). Post-industrial landscape scars are marks of sorrow and betrayal, of the abuse of power and latent hazards. At the same time, they bear tales of communities and dreams, of achievements and resistance. In short, the scars in the landscape, caused by industrial activity, constitute the flipside to the history of our modern society—integrated by necessity, yet not acknowledged accordingly.

A scar is a reminder, the trace of a wound. It is often ugly and stands for the pains of the past. Spontaneously, a scar is always understood as negative. However, some bodily wounds and scars are chosen, self-inflicted or at least positively laden. Caesarian section operation scars, Mensur scars, or body ornamentation through so-called scarification, carry different meanings and connotations, but they all have one thing in common—they are physical reminders of something of at least personal significance. A scar can be a hallmark for the veteran or the fictional hero. In a similar manner, the scar on the post-industrial landscape often conveys ambiguous and complex pasts about injustice and fear, along with survival, resilience, and courage. The story of a scar never concerns indifference; the narrative potential of the scar is a possibility and a promise we ought to embrace.

The metaphor of a scar here denotes a physical reality as well as a mental one—a mine or a factory, but also the associated narratives, experiences, and memories. Furthermore, the word "scar" is

Figure 1.1 A former Soviet military shipyard in Karosta, Latvia, raises questions about imperial dreams and hard work, lost jobs and vanished communities, as well as about contamination, new economies, and aestheticizing industrial romanticism. Photo: Anna Storm, 2001.

not only a noun but also a verb, implying a process or action taken. We can scar or become scarred, and scarring can take place. There is somebody behind the existence of a scar, responsibility and choices are involved, and in a scarring process, the open wound turns into a scab on its way to finally becoming a scar. The intermediate stage of a scab signifies a situation of undefined shapes and unsettled meanings, a liminal condition that is decisive for understanding the people and the places that will appear in the following chapters. While the scar often remains ambiguous, the scab is even more open to interpretation in a multitude of ways; here struggles over hierarchies of significance become particularly overt and discernible. Some wounds remain as scabs for a long time, because there is no room for healing and recovery.

This book will focus on the tight interrelation between our understanding of significant shared pasts and the physical structures that remind us of these pasts. I will argue for the importance of recognizing these understandings and physical structures in contemporary politics and in processes of reconciliation and search for future hope. Accentuating post-industrial landscape scars as an important, yet underinvestigated category, the aim of this book is to explore the potential of the scar metaphor as an analytical tool in order to influence the contemporary heritage debate and practice, in order to draw more thorough attention to the ambiguous, painful, and damaging pasts, as well as to possible healing processes.

Temporality and Significance

The process of healing, from wound to scar, is neither linear nor automatic. Instead, as in individual psychological healing, the process may be cyclical, can happen in stages, and even demand active work. It brings difficult pasts to the fore as often as it leads away from them, and old wounds may reopen.[1] Thus, while a scar bears the potential capacity to heal, recover, and reconcile, this is not a self-evident outcome, especially since the metaphorical scar applies to processes of healing in the social, cultural, and political spheres, rather than the biological one.[2]

Furthermore, the scar metaphor offers a way to overcome the many dichotomies of change—before and after, winners and losers, progress and decline—and create integrality instead. It is organic and created on the basis of past significances entangled with present standpoints. Because of this integrative perspective, and because it acknowledges nonlinear temporalities, the scar can be regarded as an alternative to a palimpsestual approach.[3]

A palimpsest is a handwritten manuscript on parchment or papyrus, on which one text has been scraped away and replaced by a new one. This recycling practice of Antiquity and the Middle Ages was often used repeatedly in order to make use of valuable parchment, even though it was sometimes possible to discern the erased text behind the new addition. In postmodern contexts, the word has been used metaphorically to describe the different layers of significance that make up heritage.[4] New layers are continuously added as time goes by, like new texts inscribed on a scraped, cleaned parchment. However, this metaphorical use of the palimpsest is misleading if one wishes to emphasize the interconnectedness, the linked relevance, between the different layers. If heritage consists of a multitude of parallel or successive stories and perspectives, all contributing to one another and to an overall picture, the palimpsest is a poor choice of metaphor, since the palimpsest in the original sense implies that new texts are added without regard or connection to those previously erased, and thus the end result is a picture made up of fragments and additions without any mutuality or interrelated content. Besides these metaphorical complications and diverging temporal approaches, the idea of landscape scars challenges a layered understanding of heritage per se. Instead of trying to uncover and acknowledge manifold layered stories of the past, the scar metaphor suggests that heritage is an integrated and crucial part of human living with all its inherent contradictions and ambiguities. It is a perspective that attempts to discern wholeness out of complexity and divergences. This is not to say that ambiguity is always and inherently good or liberating, but rather, as human geographer Shiloh Krupar pointed out, that "ambiguity is what must be explored to contest routinized recitations of evidence and established truths."[5] Even if the scar itself could be understood as dead, nonsensory skin, where the normal skin layers have been replaced by a dense new scar tissue, it forms a part of the living body. The scar is not a palimpsest.

Scars, Wounds, and Related Metaphors

Scars, wounds and other organic metaphors have been used in heritage or societal contexts before, primarily to describe large-scale transformations of natural landscapes, like hydropower plant dams, or traces of painful events like bombings.[6] There are also concepts that resemble the idea of landscape scars, but with a more explicit indication of the fact that affluence inevitably comes at a cost, that is, concepts that denote something exclusively negative. Among these could be mentioned "sacrifice zones," environments that have been deliberately damaged for industrial purposes in order to generate economic profit elsewhere; and "shadowed ground," landscapes reminiscent of events of violence and tragedy.[7]

The empirical characteristics associated with sacrifice zones, shadowed grounds, and several previous uses of the scar as a metaphor, such as negative environmental impacts and invisibility in wider public imaginations, also apply to post-industrial scars. However, while the scar metaphor used in this book certainly emphasizes pains of the past, this pain can be mixed and intertwined with bright, positive experiences.

The metaphors of scars and wounds have been employed by architects and artists dealing with reconstruction or memorial projects. The Swedish artist Jonas Dahlberg, for example, won the competition to design a public memorial commemorating the 2011 Utøya massacre in Norway. The general idea of Dahlberg's proposal is to create a wound or a cut in nature by taking away a "slice" of a narrow cape at Utøya. This three-and-a-half-meter void, or artificial sound, will make it impossible to reach the end of the cape; it will interrupt visitors' movement in order to "acknowledge what is forever irreplaceable."[8]

Another example is the American architect Lebbeus Woods, who has designed several projects for places marked by different kinds of crisis. His best-known proposals are for Sarajevo after the Balkan War, a project to combat Havana's deterioration after decades of an ongoing trade embargo, and for San Francisco after the 1989 earthquake.[9] His overall approach in these three places is termed "radical reconstruction," emanating from proposals called "scabs," "tissues," and "scars."[10] In many ways the architecture does resemble organic texture and form; it interplays with existing—damaged—buildings and sites, by means of contrast, mirroring, and outgrowth. In an essay, Woods articulates his ideas about this set of metaphors. He denotes a scab as "a first layer of reconstruction, shielding an exposed interior space or void, protecting it during its transformation." A scar is a "deeper level of reconstruction that fuses the new and the old... a mark of pride and of honor, both for what has been lost and what has been gained. It cannot be erased... To accept a scar is to accept existence."[11] In line with Woods' description, I believe the scar metaphor highlights a potential process toward the reconciliation or acceptance of past events, closely related to perspectives developed within the field of heritage.

Heritage and Scars

The concept of heritage is often used to signify something, usually an object, that our predecessors consciously passed on to us, or that happened to be left over from bygone days. It is also sometimes used to focus on a gift or a burden that we carry, light-hearted and proud or with difficulty and shame, to potentially hand over to our descendants. Lately, however, this conceptualization of the heritage "object" has proved to be insufficient to capture the complex interactions between humans and their physical artifacts and surroundings. Many scholars like, for example, human geographer Brian J. Graham and heritage scholar Peter Howard consider heritage as continuously changing, as an activity and a process centered primarily in the human mind rather than in the material world.[12] Heritage is defined and redefined by the present-day articulation and negotiation of values—connected to stories, places, actions, and events. This implies that heritage is a highly time-subjective activity, drawing on the past and affected by tradition and customs, but nevertheless firmly based in the current situation.

In general, heritage tends to be understood primarily as something positive. Heritage processes are seen as enriching our lives "with depth and purpose" or even to partly replace religious longings for immortality in secular societies.[13] In the best of worlds, heritage is conceived as providing a form for critical engagement in society, a way to give perceptions and identities meaning.[14] However, "difficult" or "dark" heritage, for example, in connection to postconflict situations, has recently been recognized and analyzed as well, among others by anthropologist Sharon Macdonald, geographer Dietrich Soyez, and heritage scholars William Stewart Logan, Keir Reeves, and John Giblin.[15] In fact, the understanding of heritage as something only positive has been complemented by a perspective that regards it as inherently conflictual or dissonant, since the heritage process most often defines groups of people as superior or secondary, as victims or offenders; indeed, sometimes even new conflicts arise as a consequence of the narrative selection implied by heritage.[16] In line with Giblin, I concur with the view that heritage essentially represents something neither positive nor negative, but forms an element and expression of intensified cultural negotiation.[17]

The argument of this book—to better articulate ambiguous pasts and possible processes of healing within a heritage discourse—thus connects well to the work of other scholars, but it proposes a slight change in emphasis and perspective. The empirical focus on post-industrial landscape scars makes clear that crucial memory and heritage processes can take place without any professional heritage actors involved, with the possible consequence that these processes may not be acknowledged as heritage at all. Also, similar processes can be regarded in completely different ways depending on the geographical and socioeconomic context. What is considered heritage from a Western point of view might not be laden with any such connotations at all in another part of the world, or in places where the necessities of life occupy the center of attention (figure 1.2).[18]

To grasp such processes, most of which are not recognized by the heritage profession, and to examine how they relate to already established understandings, I propose three categories of post-industrial landscape scars, representing different degrees of connection to contemporary heritage discourses. The first category, *reused* post-industrial landscape scars, is the one most closely linked to a

Figure 1.2 A team of workmen is on its way back from a shift. Past and present are entangled in many ways. Technology still in use at this steel works in Chusovoi, Russia, was consigned to the museum half a century ago in some other countries. Photo: Anna Storm, 2003.

canonized understanding of heritage. It denotes former industrial sites being redefined and reused for new purposes. The second category, *ruined* post-industrial landscape scars, denotes abandoned and decaying industrial sites, at once romanticized and considered a disgrace for modern society. The third category I call *undefined* post-industrial landscape scars, signifying places and processes that are not acknowledged as important from a memory or heritage perspective, that is, the ones that are left outside the arena of contemporary heritage recognition.

The three categories relate to one another in various ways. The reused and the ruined post-industrial landscape scars work as references and imaginary screens toward which I will try to capture the essence of the undefined scars. In some respect, the undefined scars are marked by a lack of identity and an integral potential to gravitate toward one or both of the other two categories. Some undefined scars are gradually turning into reused ones, reincorporated in the organized parts of cities and towns. Others become ruined and romanticized as a consequence of neglect. Toward the end of this introductory chapter I will describe and contextualize the three categories of post-industrial scars in more detail.

In addition to emphasizing heritage processes taking place outside professional recognition, the idea of the scar challenges understandings of heritage as regards the relationship between the mental and the physical. A scar is something you live with. It is a bodily experience, a physical memory. If shown, it is also evocative and can trigger narration: Where did you get that one? As such it can work as a crucial part of a heritage process, resembling what heritage scholar Laurajane Smith describes as a heritage "tool" or "facilitator."[19] However, the terms "tool" and "facilitator" denote only the physical artifacts or surroundings that affect heritage processes. Therefore, to describe something as a "tool" for a heritage process is to pinpoint its secondary function in the "actual" process—which takes place in the human mind. A scar, on the other hand, is an organic metaphor. It is not a tool for human beings, but an integrated part of human experience. It is both physical and mental, and by using this metaphor I intend to accentuate the closeness and mutuality between these two aspects.

Others have argued along similar lines, for example, by describing storytelling as a material practice, by asserting that places are in fact processes and moments in a network of social relations and understandings, or by claiming that buildings could be "incorporated in an urban space like a narrative in an inter-textual environment."[20] From the methodological perspective of anthropology, Amiria J. M. Henare, Martin Holbraad, and Sari Wastell suggest that we should treat meaning and thing as one identity.[21] By engaging with things, by experiencing things in a conceptual way and through metaphorical "vision," we might be able to reach beyond predefined categories of understanding.[22] Memories need footholds, and the scar metaphor provides a conceptual tool to capture both the memories and the footholds in a cohesive way.

Post-Industrial Landscapes

The notion of a "post-industrial" society gained traction in the 1970s, mainly through sociologist Daniel Bell's work, and was instantly debated.[23] The notion suggests the emergence of a new era, marked by a major transition in economies from the production of goods to the provision of services. Both the general relevance and the proposed chronological setting of the post-industrial era have been widely disputed, but one crucial turning point often mentioned is the 1973 oil crisis and its economic and political consequences. The key question was not whether the 1970s brought thorough changes, but rather whether or not these changes indicated a clear break. Many have emphasized the long lines and continuities, while others, like sociologist Zygmunt Bauman, have claimed that a definitive break in history occurred. Bauman describes how people who believed they were "forever settled" in a place—be it in geography, in society or in life—suddenly woke up to find that place no longer existing or accommodating: a situation where neat streets "turn mean, factories vanish together with jobs, skills no longer find buyers, knowledge turns into ignorance, professional experience becomes liability, secure networks of relations fall apart and foul the place with putrid waste."[24]

In the following chapters, we will explore both experiences of such dramatic changes and persisting structures that suggest societal continuity and even resilience. Post-industrial landscape scars are definitely connected with an understanding of a post-industrial society at large, but the crucial point here is not to assert that these scars are evidence of a post-industrial Western society, but to highlight people and places experiencing a post-industrial *situation*, in locations where industrial production once happened but no longer does. The scars represent specific places and communities, forming an often overlooked category in our contemporary society—whatever label we would like to assign to this society.

The word "landscape" has its etymological origin in the Germanic languages. While "land" referred to a region, territory, or environment, the suffix "scape" referred first to creation, and later, from the seventeenth century, to scenery.[25] The two parts of the word remain. We can understand landscape both as "material and territorial entity" and as a culturally defined perception of this entity.[26] While the material aspect, among other things, acknowledged a perspective of "being in the landscape" as a physical interaction of people and things, the visual emphasis brought "landscape" close to the world of art and imagination.[27]

Another way to approach the concept of "landscape" is to relate it to "land." Both of these concepts signify places full of meanings, but in different ways. Sociologist John Urry has denoted "land" as places based in the everyday, in production of goods and in the understanding of home, while "landscape" in his view is the same piece of land, also full of meaning, albeit with other connotations. Urry has showed how lands were turned into landscapes, into places of visual desire and emotion, of travel and new encounters. His examples are gleaned primarily from the agrarian countryside in the early twentieth century that became "postcarded" through the emergence of the tourist's gaze, with the technology of photography playing a key role.[28] I would argue that a similar process has also turned the industrial "land" of home and production into post-industrial "landscapes," defined by a touristic visual experience of abandonment, rust, and growing vegetation. Urry asserts that the transformation from land to landscape is irreversible, but I propose that landscape scars can indeed form a reconnection between the two, between land and landscape. The scar might be the result of this transformation, but it is also a creative possibility, a new entity providing integration between different stories and significances. Hence, the experiences and perceptions of land and landscape can exist at the same time, at the same physical spot, but the constitutions differ and one of them is dominant, either the land or the landscape. The scar is a possibility to acknowledge the abiding meanings of land within an understanding dominated by landscape. People living in the land, who consider it their home, probably know whether it has become "postcarded" into a landscape by other people and vice versa: people comprehending a place mainly through the tourist's gaze can include an understanding of a still existing land.

Industrial Heritage

Heritage is not only a perspective and an analytical framework. It also signifies a kind of activity with its own history. My interest in post-industrial landscape scars has not emerged in a vacuum, but has been influenced by at least a century of public and individual concerns for industrial remains. In the early years of industrialization, industry was in many respects seen as the opposite of culture. Heritage, on the other hand, was perceived as a condensed expression of culture, and consequently it was difficult to comprehend industry as something belonging to a heritage process.[29] Industry represented novelty and innovation and hence almost an antithesis to the past. Instead, industrialization triggered a heritage interest for agricultural milieus.

However, at the turn of the twentieth century a conscious defining of some industrial techniques and built environments as belonging to the past had already emerged.[30] During this period, the function of the identified industrial past, among other things, was to work as a contrast to modern industry and modern society as a whole. This was also a time when Western countries established institutions that were to deal with the past in more general terms, in order to form a foundation for modern society. Industrial remains were acknowledged as possessing heritage values at some level, and included in exhibitions, documentation projects, and museums.[31] The initiative came from industrialists in league with academic scholars, that is, representatives of an elite in society. A major shift in scale and perspective took place in the 1960s and 1970s, with a general Western questioning of society. Rapid changes in the industrial landscape and structural crises contributed to a greater awareness of, and interest in, the past.[32] Common denominators of the new heritage activities were an emphasis on everybody's right to take part in the writing of history, and an understanding of heritage as having social and political implications. In short, the activities of the 1960s and 1970s expressed a view "from below," where workers were considered experts on their work and therefore best suited to write the history of that work.[33]

Consequently, the workers' situation, detailed technical processes, and the local community were put at the center of attention, which contrasted sharply with the conventional heritage processes and narratives of the time.[34] Industrial archeology in Great Britain and the United States, "eco museums" in France, and the so-called dig-where-you-stand movement in Sweden were all some of the expressions of the radical spirit of this time.[35] In many of these activities there was an inherent tension between academics and amateurs, and between, on the one hand, those who regarded knowledge about industry, technology, and industrial architecture as the prime target of their work and, on the other hand, those who focused on the social dimension of the former work sites.[36]

In the following decades industry was gradually included in the general understanding of heritage. Today industrial heritage draws great interest both from scholars and the general public. However, industry has not become the core of canonized high status heritage when it comes to heritage management practice.[37] There are many reasons for the hesitancy to include industrial memories in an "authorized heritage discourse," to use Laurajane Smith's concept, among them the fact that built environments related to an industrial past are often of a mundane character, consisting of complicated, large-scale, polluted, or otherwise devastated landscapes.[38] Post-industrial landscape scars are often too dilapidated, or too commodified, or too complicated to be easily recognized within a heritage perspective. The scars might also have been too stigmatized to be included in an everyday landscape. Some sites or events may, it has been suggested, "remain scarred indefinitely" implying that healing would be impossible.[39]

In spite of the difficulties, these leftovers or side effects of industrial production are not only crucial for a richer understanding of the many faces of industry, but, as I argue in this book, they form scars in the landscape that convey narratives of resistance and recovery of a much wider societal significance. The physical aspects are prominent in terms of, for example, contamination, affected ecosystems, and townscapes, as are the mental aspects in terms of processes of heritage and identity. To heal a mental or physical wound into a scar that one can live with is to recognize key signs of difficult or ambiguous pasts and to point toward possible reconciliation. Although the dividing line between healing on the one hand, and obscuration and inequities on the other hand, is thin and requires a careful balancing act; the scar can indeed offer a focal point for meaning and depth in our lived post-industrial landscapes.

Localized Utopian Visions and Conflict Zones (the Places of the Book)

In the following chapters, I will show you some scars in our post-industrial landscape. The geographical setting is the Baltic Sea Region of Northern Europe: this study will focus on people and places in Lithuania, Germany, Denmark, and Sweden (map 1.1). The histories of these scars all speak of twentieth-century utopian visions of society, of fear and resistance expressed by popular movements, of individual and state investments of considerable dimensions, and of special relationships between industrial workers and those in power. The scars also highlight how big plans become reality in a local context, and how the affected local context remains when the big plans sometimes vanish.

In my opinion all of these places are remarkable, spectacular, and astonishing. At the same time, they are generally regarded as peripheral, ugly, and merely functional or dysfunctional—if they are noticed or known at all. Why is this? This has been a personal starting point for my investigations. How come these obviously significant narratives have not been written? Why are these sites not full of tourists, these scars not recognized? One part of the answer is connected to limited accessibility and to risk, but that certainly is not the entire explanation.

Chapter 2 explores what it means to live on an *unstable mountain* in the mining town of Malmberget in the far north of Sweden. Here, the mine is the sole reason for the community's existence and has brought work and welfare as well as pain to the inhabitants. The town has flourished and suffered because of the mine and, since fifty years back, one feature above all others epitomizes this relation: the Pit.

INTRODUCTION

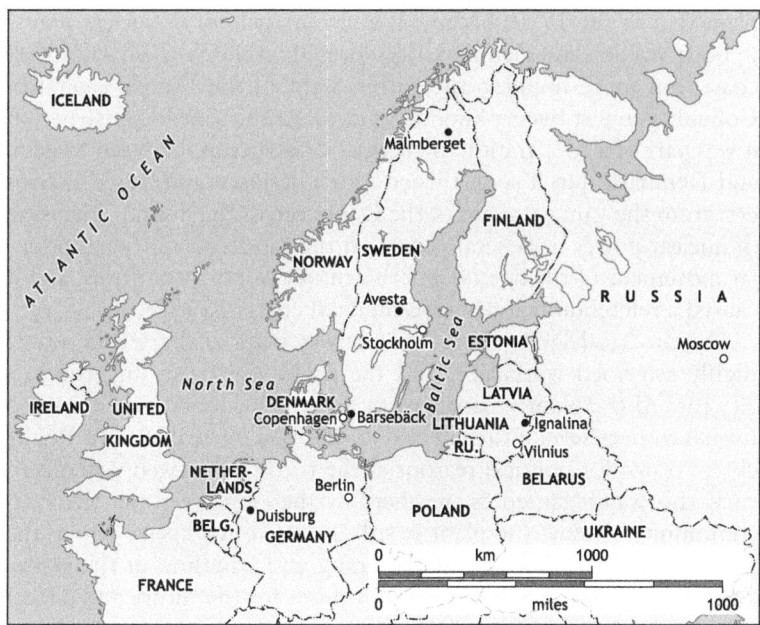

Map 1.1 The post-industrial landscape scars of this book are located in the Baltic Sea Region, in the countries of Lithuania, Germany, Denmark, and Sweden. Map: Stig Söderlind, 2013.

The Pit is a huge hole—200 meters deep and comprising an area of 21 hectares. Originally it was an open-pit mine that later continued to grow because of ongoing underground mining activities. Today it divides the town center into two halves—people have been forced to abandon their homes and key community buildings successively because "the Pit is coming."

In this dramatic, yet slow, process, the mining company is surprisingly silent. The concerns of the local population, the miners and their families, are voiced but the responses are few. The contrast to the nearby town of Kiruna—where the same mining company is planning to move the whole town because of the needs of the mine, in a highly medial spectacle—is striking. The scar in Malmberget is the literally disappearing homes and places of childhood in the midst of a dying town, along with a negligent attitude from the mining company toward a loyal, yet hurt, local community.

Chapter 3 asserts that there exists a geographical *distance of fear* by exploring the history of Barsebäck, a commercial nuclear power plant that in the 1970s became the utmost symbol of nuclear power in both Sweden and Denmark. It is located on the Swedish southwest coast, but more importantly, within sight of the Danish capital of Copenhagen just twenty kilometers away. At the time of construction it was part of a cooperation in energy infrastructure between Sweden and Denmark, but it soon turned into a despised and feared silhouette from the vantage point of the Danes across the Sound. The issue of nuclear power in general generated the most encompassing popular movement of the late twentieth century in both countries, and it caused a referendum and a governmental crisis in Sweden.

On the local level, the nuclear power plant of Barsebäck was a highly esteemed workplace, and the pride, the trust, and the loyalty toward the plant's management and the nuclear sector remained unshaken over several turbulent decades. When the plant was finally closed down for political reasons at the turn of the twenty-first century, this was regarded as treachery by the employees and the local community. Today, the plant is still an active workplace where the task is to prepare the decommissioning and handling of the waste. The scar in Barsebäck marks lost optimism for the future and a local experience of being betrayed, the heated controversies on nuclear power in Sweden and Denmark, and not least, the uncertain future of radioactive waste.

Chapter 4 examines a *lost utopia* in connection with the Ignalina nuclear power plant. This plant was established in the context of the Soviet empire and its large-scale ambitions for electrification. In the 1970s, the Soviet republic of Lithuania was chosen to host the largest nuclear power plant of its kind in the world. A few kilometers away, Sniečkus, a workers' town for thirty thousand inhabitants was built. Because of the industrial enterprise, the town was semiclosed and received special benefits. It was built according to contemporary planning ideals and regarded as a "Soviet paradise." Russian speakers from other parts of the empire moved to Sniečkus, while the presence of ethnic Lithuanians was negligible.

Over just a few decades, the Ignalina nuclear power plant made a remarkable symbolic journey from being an expression of advanced Soviet technological progress, to becoming a despised sign of Moscow's imperialistic ambitions threatening human health, the environment, and ultimately the Lithuanian nation, only to emerge in the 1990s as a most valuable energy resource for independent Lithuania in its relations with Russia and the West, and then ultimately ending its operational life as a concession to negotiations for Lithuanian EU membership at the beginning of the twenty-first century. The scar of the Ignalina nuclear power plant is the experience of a lost utopia, prevailing and poisoning ethnic tensions and, again, the unavoidable and hazardous handling of radioactive waste.

Chapter 5 deploys the idea of *industrial nature* by visiting an industrial area with similarly high symbolic overtones to its country: the Ruhr area in Germany. This heavily industrialized conglomerate of cities and towns has a complicated history as the national armorer's workshop during both world wars, besides severe industrial pollution, heavy work, and an extensive immigration situation that brought about social tension. In the 1970s, the area faced a severe economic crisis when many of the industrial activities ceased—with huge unemployment as one consequence. Massive programs to bring new life to the area have been carried out since, perhaps the best-known of which was the Internationale Bauausstellung (IBA) Emscher Park, 1989–1999. Artistic interpretations, historical investigations, and adventurous reuse of the abandoned industrial structures characterized the investments.

One of the landmarks was the Landschaftspark Duisburg-Nord, where different parts of a former blast furnace plant were reused for diving, climbing, biking, concerts, and exhibitions, among other things. The industrial structures were converted into frames or sculptures for types of activities that are usually more connected to a natural environment. A special understanding of industrial nature also developed, with controlled overgrowing becoming a strategy to revive the contaminated sites. The scar of the Landschaftspark Duisburg-Nord is the complicated history of war, contaminated ground, and a partly lost industrial identity, along with difficulties of long-term hope for the future once the initial enthusiasm for the spectacular changes has faded.

Chapter 6, finally, scrutinizes the *enduring spirit* of the company town Avesta in the heartland of the Swedish iron and steel industry. In Avesta there was a copper works and later an iron and steel works that used the power from a river and formed the basis of a company town for centuries. In the mid-nineteenth century, the original industrial site became too small for the needs of the growing production, so a new industrial area was established on the other side of the town and separated from the river, based on the transmission of electric energy rather than water power.

The old site was mostly abandoned and stood more or less empty behind fences for a few decades. Then, newcomers to the town found the abandoned ironworks and turned it into an art exhibition hall, followed by other projects like restaurants, a museum, business offices, and school facilities. The area was opened up to the public for the first time and the reuse program was recognized as interesting and of high quality from a national perspective. As in Duisburg in the Ruhr region, the scar in Avesta thus mainly belongs to the reused category. Yet this mostly healed wound also shows difficult and chafing aspects, like the conversion of the area from a place for workers to a gentrified elite location, a place where the former ironworks employees never go.

Although both Duisburg and Avesta can be regarded as reused landscape scars, combined with some ruination, a big share of this book will be dedicated to the undefined scars. Often little appreciated and "invisible," these scars bring heritage matters to a head. Who is a legitimate interpreter? What is the effect of raising questions from the outside, for example, as I do in this book? Is there a risk of losing important parts of the experience during the healing process? In order to frame how undefined scars may gravitate toward reuse or ruin, in this final section I will expand a bit on these latter categories.

Categories of Post-Industrial Landscape Scars

The *reused* post-industrial landscape scar is in its physical sense typically a nineteenth-century brick building with large windows located along the waterfront in the center of a city or town. It is old enough to be regarded as beautiful by a large proportion of the public and it is reused for housing, exhibitions, restaurants, schools, or offices.[40]

It is possible to argue that the history of the reused post-industrial landscape scar began in the late 1960s. The post–World War II decades, dominated by new construction, were followed by an interest in adaptive reuse.[41] The new approach put the reuse of abandoned factories on an agenda of economical and sustainable behavior, along with an appreciation of cultural and historical values.[42] Spurred on by initial projects in the United States, particularly the development of the waterfront in Baltimore, Maryland, the renovation of buildings in the textile city of Lowell, Massachusetts, and the 1970s loft living movement in New York, the reuse of abandoned industrial places became prominent in Western Europe during the 1980s and 1990s.[43] The contrast between "before" and "after" is often emphasized in the depiction of these reuse processes. Lowell, for example, has been described as a city that turned from depressed mill town into a "vibrant, revitalized and healthy" community by transforming the vast amount of vacant mill buildings into premises for high-technology companies.[44]

Today industrial features are definitely an asset in many contexts. It is easy to find commercial advertising drawing on an industrial past as a marketing tool. For example, potential buyers of apartments in a former mill in Gothenburg in Sweden are enticed by a lifestyle characterized as "warehouse living," a concept said to bring qualities from the past into an encounter with the present and thereby "create a unique living environment."[45] This unique living environment comprises, among other things, preserved visible steel girders and plastered interior walls, original deep windowsills and open floor spaces—elements the rather wealthy buyers apparently favor.[46] An industrial past, expressed in prominent material features, is also a benefit and key feature in many prestigious landmark projects, like the Bankside Power Station in London, which opened as the art museum Tate Modern in the year 2000. The attention toward industrial aesthetics within the reused category has extended so far as to trigger the construction of pure replicas or "fake" industrial structures to help sell apartments and to stage spectacular shows.[47]

The reused scar is often characterized by gentrification, a status upgrading that includes new tenants as well as new uses.[48] The industrial past is adapted to fit into a new narrative of upper or middle-class life, flavored with what can be regarded as interesting industrial details that bring an experience of authenticity to the new inhabitants.[49] However, the appreciation of the industrial features also brought a new overall framing to many places. The industrial character has often been contrasted to, or combined with, an enhanced "natural" setting, especially visible in the conscious addition of greenery and in processes of renaming. Hence, one can easily find former industrial places behind names alluding to the sea, the shore, and the valley. A waterfront location, once a decisive factor as the source of energy, has been turned into an attribute for living space quality.[50] As a consequence, the industrial aesthetic and the industrial past have become both commodified, and—in the words of Sharon Zukin—"domesticated."[51]

The early reuse and commodification of post-industrial landscape scars did not involve many heritage professionals, many of whom questioned the process.[52] For example, industrial heritage scholar Neil Cossons rhetorically asked whether sites that "survived the years of neglect [will] now survive the period of rampant rehabilitation?"[53] The reused landscape scar was commodified and domesticated, made visible, agreeable, and respectable, and thus, from the perspective of heritage professionals, sometimes healed in a bad way, concerning both its physical and its mental structure. In general, the places have been cleaned and the stories adapted for those who do not have personal memories of the place when it was in operation in its previous use, although there are, of course, exceptions.

It seems as if heritage and planning professionals to some extent found a common platform in the mid-1980s when, on the one hand, heritage rhetoric began to be used to advertise offices and apartments, and, on the other hand, planning and development ambitions were used to justify the existence of heritage activities in society, not the least in terms of industrial heritage tourism. These activities range from UNESCO designating industrial sites as "world heritage," to the advertising of industrial heritage as an exploration into the exciting and unknown. In this respect, the reused post-industrial landscape scar closely connects to the ruined scar, or at least it becomes clear that they belong to the same family of marks in the landscape.

* * *

The *ruined* post-industrial landscape scar is in its physical sense typically a twentieth-century factory or infrastructure establishment like offices or canteens or a power plant, located in the countryside where the pressure of land development is low or zero. These are places where people have left their binders on their desks to never return, and where shrubs grow through the broken windows.

Ruins in general have been appreciated, even cherished, since the late-eighteenth-century Romantic period. In paintings, poetry, and as fake ruins in private gardens, the ruin was established as an emblematic symbol of the cycle of life and death, and the inevitability of time passing—in short a prime ingredient of the Romantic scenery. A patina of age sent signals of authenticity and was considered aesthetic.[54] Two hundred years later, in the late twentieth century, came what human geographer Tim Edensor has called "the golden age of industrial ruination."[55] In contrast to the praised ruins of the eighteenth and nineteenth centuries, the industrial ruin did not immediately evoke similar connotations of melancholic beauty, but instead represented a wasteland of dark urban nightscapes and abandoned parking lots that were loaded with meanings of ugliness and danger. After some time, a romantic, mysterious, and tempting aura of exotic otherness was nevertheless established in connection with these places as well. The emergence of concepts and activities like "urban exploration" and "industrial cool," on the fringe of official heritage actions, show how common and accepted these interactions with the ruined post-industrial landscape scars have become.[56] The visual experience definitely dominates the ruined scars. Amateurs and professionals alike indulge in what has become a photographic genre of its own. The image of material decay and the nature "taking back" the human-shaped environment is a popular motif in blogs as well as in coffee table books, and industrial ruins have also been used as stages in countless films.[57] It has been argued that abandoned industrial sites form necessary counterareas within an all too well organized contemporary Western society. As such, they can question normative regimes of memory and materiality.[58] Yet it has also been argued that decaying urban-industrial environments have been romanticized by photographic depictions to the extent that past and present social injustices are concealed, for example, in Detroit, the largest modern US city in ruins, with currently as many as eighty thousand abandoned buildings.[59]

Hence, the visual aspect is crucial and also forms part of the explanation as to why the undefined post-industrial landscape scars are not recognized—they are simply not seen. And why? The reason could be that they are located in places difficult to visit or comprehend, or that these scars just do not stand out as important.[60] The time perspective is also decisive, since both the reused and the ruined landscape scar categories are definitely in a post-industrial situation, while the undefined scars sometimes remain in a liminal and uncertain position, which makes it complicated both to look back to the past and forward to the future.

The experience of the post-industrial landscape scars could thus be intellectual, visual, and tactile, understood in some respects and a complete mystery in others. A poem by the Swedish writer and journalist Göran Greider published in 1995 captures this ambiguity:

> The factories though remained forbidden cities.
> Childhoods passed by in the sign of a mystery.
> The adults were averted, inaccessible.
> When the factories become silent they turn visible again.
> Not until now they are unfamiliar to us.[61]

The factory has turned visible by becoming redundant and then being rediscovered, but to some extent this change remains incomprehensible. Post-industrial landscape scars carry elements of loss, but also emerging new understandings that fill the emptiness with significance.

Chapter 2
Unstable Mountain

Ore is an economic term. Rock is labeled ore only if it is worth mining, which depends on the market price in combination with the costs for mining and processing. In many mining towns, the economic rationale is generally predominant, for the mining companies, for the miners, and for the way the built environment is handled. Nevertheless, other kinds of values do compete in the continuous process of defining the landscapes and activities marked by mining. The economic logic is challenged by claims of significance connected to, for example, tradition and hope for the future, professional pride, perceptions of beauty, and places of home. So far, however, economic rationale has been supreme in the mining business, and deafness to other viewpoints has scarred both people and landscapes. Open pits, underground shafts, and galleries can never be undone, and some mountains will never be stable again.

Among the different kinds of scars mentioned in the introduction was body ornamentation called "scarification," an extension to the idea of the tattoo. It is a painful procedure to cut or burn patterns in the skin, indeed to self-inflict a wound. The healing of the wound is often consciously delayed by irritation so as to produce a distinct scar, red to begin with and later turning whiter in appearance. During the healing process, the scar usually spreads or moves and any details in the pattern are lost. Some of the characteristics of the practice of scarification relate to the scarring conveyed by mining activities: they are both consciously chosen and their position is not fixed for all time. A decisive difference, though, is that in scarification the scar itself constitutes the goal of the activity, whereas in mining the main objective is to make profit from mineral extraction.

In this chapter, we will explore the moves of a consciously chosen but undesired post-industrial landscape scar. The main focus is on a

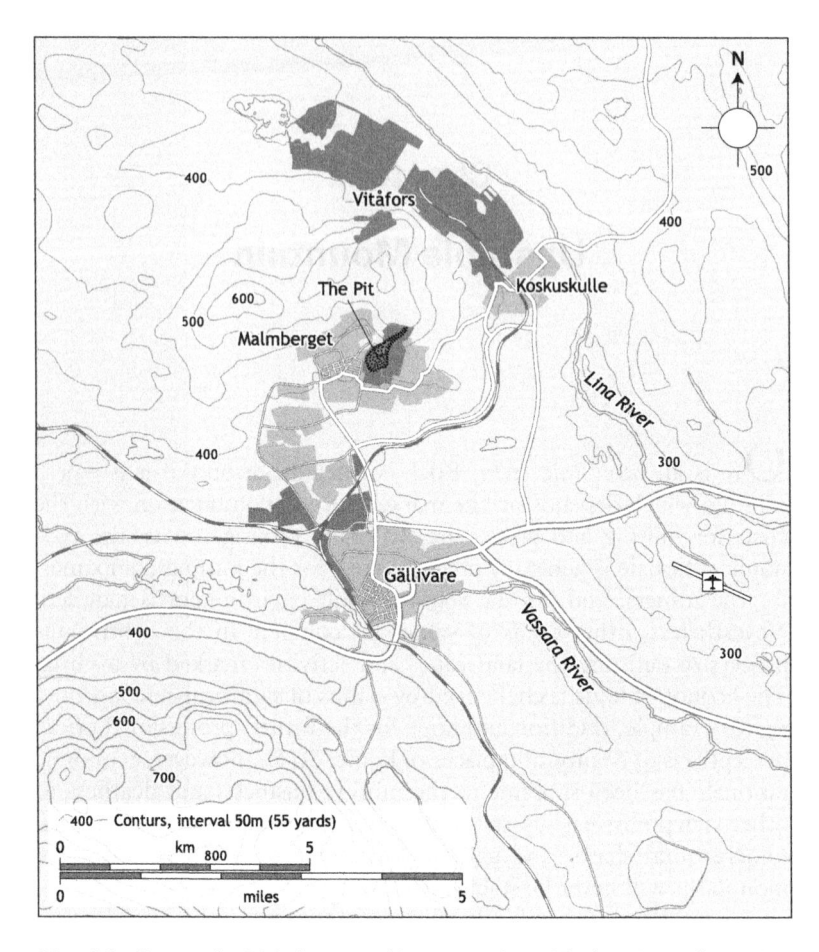

Map 2.1 Present-day Malmberget and its surroundings. The location and extension of "the Pit" is marked in the middle of the town center. South of Malmberget is its twin town Gällivare, and between them is a new residential area, Mellanområdet. North of Malmberget is the industrial area of Vitåfors, where aboveground mining activities are centralized. Map: Stig Söderlind, 2013.

continuously growing open pit in the mining town of Malmberget, Sweden, which forces the inhabitants to relocate or tear down their homes and key public buildings, to abandon the land and find new places to live (map 2.1). The open pit is part of the mine that forms the exclusive basis for the town's existence, and thus, the pit simultaneously "feeds" and "eats" the town. I will also compare Malmberget with nearby Kiruna, another mining town forced to move.

Rich Mountains of the North

The mountainous area of northern Sweden was long inhabited mainly by the indigenous nomadic Sami, who used it for grazing reindeer. In the seventeenth century, a mountain called Gällivare malmberg was identified as rich in iron ore—"malmberg" simply meaning "ore mountain"—and in the eighteenth century small-scale mining was started, headed by several subsequent prospectors. At the beginning of the nineteenth century, the estimations of the amount of iron ore in Gällivare malmberg made by the national steel producers' association board soared: "[It is] most probable that the mountain itself constitutes one single continuous mass of iron"—although it later became apparent that the iron ore was spread out over several separate ore bodies.[1] The main obstacles toward larger-scale mining were the difficulty of transporting the ore to the processing blast furnaces along the coast, as well as technical issues concerning the purification of the phosphorous ore. With the breakthrough of the so-called Thomas process in the 1870s, large-scale steel production based on phosphorous ore became possible, and in 1888 the railway reached Gällivare malmberg. Together, these events drastically changed the prerequisites for mining at the site, and the area soon became Sweden's Klondike.[2]

Even as the railway was being built, the sparsely populated area suddenly filled with newcomers. In 1888, the trade and church village of Gällivare had less than three hundred inhabitants; the railway brought it more than three thousand workers.[3] When mining began, there was far from sufficient housing, and since the earnings in Gällivare malmberg were double or triple what was paid for corresponding jobs in the mines of southern Sweden, fortune hunters came from far and wide.[4] Most of the little housing that did exist was located in Gällivare, and the five kilometers between the village of Gällivare and the ore mountain with the actual workplaces were too far considering the modes of transportation at the time. As a result, a shanty town appeared directly adjacent to the mine on the mountain.

The living conditions in the shanty town were slum-like, and contemporary observers expressed their dismay over the level of alcohol consumption and prostitution. People lived in soil huts or simple constructions they had built themselves from empty dynamite boxes and the like. Initially, British-owned companies headed the mining activities, but after a bankruptcy a Swedish-owned company was formed in 1891, AB Gellivare Malmfält (AGM). The mining management saw the need for more regulated housing, so the County Administrative Board hired an engineer to make a town plan. However, conflicting interests protracted the process considerably. In 1899, a town plan was eventually approved for the settlement now officially called "Malmberget"; during the first years of the twentieth century this regulated, built environment expanded rapidly on the mountain slope while the shanty town was successively dismantled.[5]

In the same year, 1899, the railway headed further toward the mountains of Luossavaara and Kiirunavaara about 120 kilometers north of Gällivare and Malmberget. In this area, too, rich iron ore deposits had been discovered years before. However, the situation in Malmberget had made national mining authorities reluctant to grant permission for another mining site in fear of repeating the negative experiences of an uncontrolled shanty town.[6] There were also considerations about how far north the railway should be built, bearing in mind potential military exposure toward Sweden's long-standing enemy, Russia. The mining company that applied for the mining concession had to assure that it would immediately build proper housing for the workers in what was to become the town of "Kiruna," and the Swedish Parliament decided to finance a substantial fortress at Boden, located closer to the coast, to defend the new national treasures of the north. In 1902, the railway stretched from Luleå on the Swedish east coast to Narvik on the Norwegian west coast, passing the mining areas of Malmberget and Kiruna on its way. The new railway was built to transport ore from the mines to the shipping ports in Luleå and Narvik, this purpose being obvious not only in its location, but also in the technical equipment with which it was outfitted. For example, the inclination of the tracks was low while the maximum load weight was very high to support the heavy ore transports.

The mining company in Kiruna was named after the two mountains Luossavaara and Kiirunavaara and the Swedish abbreviation for limited company, AB, forming the "LKAB." After just a few years, the mining company AGM in Malmberget became the majority owner in LKAB and soon merged its activities with the affiliate. At the same time, the Swedish state entered LKAB as a shareholder and, in a private-public partnership, LKAB thereafter carried out mining in both Malmberget and Kiruna, being the undisputed dominant player in the new industrial sector. The mining business in northern Sweden gave rise to what has later been termed a "technological mega system" including the mines, the railway, shipping ports, and hydro-power plants.[7]

While the town plan in Malmberget was intended mostly to clear the slum and encourage decent housing, the ambitions for Kiruna aimed very high indeed. The company engaged a number of the best-known architects and planners at the time, including names like Gustaf Wickman and Per Olof Hallman. Under the leadership of the dynamic LKAB manager Hjalmar Lundbohm, Kiruna developed into a "model town" in the sense of high-quality architecture and modern infrastructure adapted to the terrain and the harsh local climate, but also including a pioneering school system and a lively civil society.[8] In both Kiruna and Malmberget the mining company built large parts of the towns, along with areas built and managed by the state railway company. A third actor, the respective municipalities, gradually entered the arena of building construction and completed a structure of three distinct built-up areas in both towns—the so-called company land, the railway area, and the town plan area. Around 1900, Malmberget had about seven thousand inhabitants and Kiruna about two thousand.[9] The Malmberget twin town relationship with nearby Gällivare had no equivalent in Kiruna, a circumstance that would prove important later on.

In Malmberget the mine was based on about 20 different ore bodies, located in an area five kilometers in length and two kilometers wide, which implied a scattered spread of mining sites.[10] The town took shape in the slope just south of the mining sites. In Kiruna, on the contrary, there were only two main ore bodies, one in Luossavaara and one in Kiirunavaara, and the town was built between the two mountains. Open pit mining was successively replaced by underground mining, first in Malmberget and then in Kiruna.

From the start, Germany was a key customer of iron ore from Malmberget and Kiruna, and during World War II this became a highly contested issue. Sweden formally remained neutral in the war. In spite of protests from the Allies, it continued to ship iron ore to Germany throughout the conflict, both via the port in Luleå and via Narvik in Nazi-occupied Norway. The Swedish rationale was to avoid the risk of a conflict with Germany, and the fact that the iron ore export brought substantial income to the state certainly did not discourage such business. The importance of Swedish iron ore supply for German warfare had a substantial impact on the course of the war's development.[11] LKAB generally leaves out this part of the company's history, although it acknowledges that the production graph of the mines "mirrors the historical development of the Ruhr industry and the whole of Germany."[12]

An "Expendable Community"

As in many other Swedish industrial branches, the postwar decades brought an economic boom in mining connected to the rebuilding efforts on the European continent. However, the flourishing mining in Malmberget also had a flipside, namely, the realization that some of the ore bodies might stretch under the town so that, if mined, the cavities would make the ground unstable and unsafe to live on. In short, the existence of the town was threatened by the ongoing underground mining activities (figure 2.1). In 1956, the County Administrative Board initiated a comprehensive planning conference to deal with the situation, and a consultant worked for almost a decade to complete a new plan for Malmberget, which at this time housed nine thousand people in what had become close to an ideal welfare community.[13]

Figure 2.1 Drilling in Malmberget around 1965. Work underground is strenuous but relatively well paid. It has long been a male-dominated world. Courtesy of Gellivare Bildarkiv.

However, the probable extension of the ore bodies had actually been known for some time. Half a century before, in connection with the first, protracted process to produce a town plan in Malmberget, the mining company stated in a letter to the county administration that it was "probable that the ore deposits now mined in the Captain Hill [Kaptenshöjden] extend underneath the eastern part of the town plan in question, such that it is possible that mining will be carried out here in the future."[14] The company also offered another location for building the regulated town, but for various reasons, among them favorable wind directions and proximity to services, the town plan of the late nineteenth century was approved on the same spot as the shanty town it was to replace, that is, the location potentially on top of the ore bodies.

Therefore, the situation should not have come as a surprise in the 1950s, but apparently it did and the debate became heated, reaching even beyond the local context. One of Sweden's daily newspapers claimed that Malmberget was in danger of being snapped in two with an "open wound" in the middle, referring to a predicted expansion of the old open pit connected to the "Captain ore body."[15] This prophecy was vigorously refuted by a geologist, who stated in the LKAB staff journal that "Malmberget will surely remain habitable to the extent that it will not be divided in two!" He admitted that LKAB had indicated that there was "the possibility of some degree of risk down toward Tingvalls Road, but also that it will probably be a hundred or so years before any landslide occurs so far down."[16]

However, a group of experts at the 1950s planning conference advocated that the best thing to do would be to move the whole town of Malmberget five kilometers south to Gällivare and to halt all new construction in Malmberget. Shortly afterward, the LKAB staff journal argued in line with this that it was important to avoid "repeating the mistakes already made in Malmberget, where the town plan was developed in such a way that necessary changes will soon be warranted."[17] At the beginning of the 1960s, LKAB CEO Arne S. Lundberg noted that large parts of the town might indeed be affected by mining activities and, like the planning experts, he envisaged a development in which Malmberget would "move in the direction of Gällivare so that the old dream of the two towns coalescing could become a reality."[18]

From its formation and up to the 1950s, Malmberget had been the obvious hub in relation to Gällivare, among other things, in terms of trade and services. Thereafter, more and more merchants chose to settle in Gällivare, due to better communications and available land, and, one might guess, the discussions about potentially unstable ground in the future Malmberget. In 1960, Malmberget and Gällivare merged administratively to a joint municipality, and there were certainly visions of the two towns becoming one, among other things, through construction of a highway between them. However, a few years later the comprehensive plan consultants commented that this wish was still "an uncertain dream for the future."[19]

The two towns were different in many ways, not only in terms of historical background and dominant identities, with Gällivare the old church village and trading post and Malmberget the modern industrial mining town. Their natural characteristics also differed. Gällivare's highest point, 375 meters above sea level, was equivalent to the lowest point in Malmberget. The difference in altitude within Malmberget was no less than 100 meters, with some roads so steep they could not be used by cars.[20] Apart from the extensive views Malmberget offered of the surrounding mountain landscape, the local climate also favored the mining town with decisively milder winter temperatures, less wind, and also—because of its higher location—more daylight during the winter time. This was indeed important, considering that Gällivare and Malmberget are located about 70 kilometers north of the Arctic Circle, such that the midnight sun in the summer is balanced by only about two hours of daylight during some weeks in the winter. In Kiruna, even further north, there is complete darkness for several winter weeks before the sun again rises above the horizon.

In the 1960s, there were thus competing conceptions of the future of the town of Malmberget. Was it to become divided in two by an open sore in its center or not? And what about the twin towns of Gällivare and Malmberget? Should they merge into one entity, or keep their distinct and separate characteristics? At this time, the amount of iron ore available in Malmberget was estimated to "far exceed" the amount in Kiruna. In spite of the unclear consequences of the underground extension of the ore, LKAB emphasized that public service was to be encouraged in Malmberget for the benefit of its employees.[21]

Ore surveys are long and drawn-out processes, ending in qualified estimations. In 1961, the LKAB CEO Arne S. Lundberg asserted that the extent of the ore bodies extending under Malmberget would probably result in mining operations "for at least 200 years ahead."[22] However, he also insisted that the sheer existence of ore deposits was not the only prerequisite for mining. In practice the financial issue was the most important factor determining the extent of mining activities, and thereby also for employment and the local economy. Lundberg accordingly pointed out that "uncertainty about the details of the town's future development will prevail."[23]

During the municipality's long work with the comprehensive plan, LKAB also produced new information about the expected number of workers in the mining industry, which forced the planners to repeatedly rethink their prognoses and suggestions. When the plan was finally presented in 1967, the planning consultants stated in the foreword that "extremely few [towns] would have presented such complicated conditions" as Gällivare and Malmberget.[24] They also emphasized that the plan should not be regarded as complete, but as a framework for rolling five-year and annual municipal action programs: "For its own sake, public enterprise must, like industrial enterprise, be engaged in continuous planning and thereby always be thoroughly prepared for new developments and essential investments."[25]

During the ongoing work with the comprehensive plan, LKAB also announced that the ore located underneath the center of Malmberget was, indeed, soon to be mined. The company began to buy privately owned buildings in the town center in order to evacuate the inhabitants and tear down the buildings. In the comprehensive plan and in the LKAB staff journal, an unsentimental relation to the existing built environment was expressed. For example, it was briefly noted that in a first step "the church, the grammar school, the old public baths, 'Folkets hus' and some shops and housing areas will have to be demolished."[26] It was considered almost a sheer technical question, and Malmberget was even characterized as an "expendable community," that is, a community that was to be used, consumed, and then discarded.[27] Many buildings in the very center of Malmberget did disappear, among them the school, "Folkets hus," and the public baths. Most of them were blown up, with the roads left unused in a leveled and fenced area. The church, however, originally a gift from LKAB to the parish of Malmberget, was moved to a new and safer location in the town.

The number of mine shafts spread throughout Malmberget had been centralized at the hoisting facility at Vitåfors in the 1950s, so that all transportation and processing of iron ore was concentrated there. As a consequence, in the early 1960s, the railway area in the center of Malmberget became redundant and was dismantled. The abandoned railway area was instead turned into a new town center, with a new school, a new "Folkets hus," a new public bath, and an indoor ice rink. In addition, a new high-rise apartment building was erected, accompanied by commercial premises.

According to newspaper reports, the reactions to the ongoing changes among the population of Malmberget were incongruous. On the one hand, the local population was upset and expressed sorrow and concern about the course of events, strongly criticizing LKAB's actions. On the other hand, it seems as if many people had suppressed or neglected the information about the threats the mining presented.[28] However, on New Year's night in 1971 a landslide occurred in the evacuated area, creating a crater 30–40 meters wide in the town center.[29] Historians Fredrik Gustavson and Pär Isling argue that this landslide finally made the residents of Malmberget and its representatives aware of what was going to happen to the town. As one consequence of this awareness, during the 1970s a new residential area began to take shape between Malmberget and Gällivare called "Mellanområdet," meaning "the area in-between."[30]

In Kiruna the postwar mining boom brought new additions to the town's collection of architectural landmarks designed by famous names, for example, an apartment block by Ralph Erskine and a town hall designed by Arthur von Schmalensee, both from the early 1960s. And although mining dominated economically in Kiruna as well, other lines of business contributed to the image of the town; these included a military regiment, the space research center Esrange established outside the town in the 1960s, and tourism to Abisko in the mountainous area, which had been designated a national park way back in 1909 and attracted tourists and hikers ever since.

* * *

During the first half of the twentieth century, many large industrial concerns in Sweden bore their own transport, repair, and maintenance costs, ran their own farms, and managed their own housing areas for their employees, where they also took responsibility for infrastructure like water and sewage disposal, fire protection, and roads.[31] LKAB was no exception. However, at the end of the 1940s many companies began withdrawing from such activities to concentrate on what they regarded as their core mission. Municipal responsibility then developed in line with this.[32] A shift of focus can also be discerned in the relations between LKAB and the municipality, although it did not appear until the late 1950s and early 1960s in connection with the forced evacuation of the central areas of Malmberget.[33] At that time, for example, LKAB ceased to produce milk on their own premises, and the company's mining school was incorporated into the municipal vocational school. In 1965, an agreement was reached in which the municipality assumed responsibility for the construction and maintenance of streets, roads, water, and sewage in certain parts of Malmberget.[34] LKAB's involvement in the Malmberget community nevertheless remained on a high level compared to other companies in similar towns.

A mutual dependency between the municipality and the mining company was manifested not only in the company taking care of water and streets, but also in the fact that LKAB and its employees provided 60 percent of the municipality's tax revenue.[35] However, there was hardly ever any simple relationship between expansion and decline in the mining operations on the one hand, and the number of jobs for the local community on the other, particularly since development in mining technology successively implied an overall reduction in the workforce.[36] Instead, as we have seen in connection with the pursuit of mining that forced the town center to be abandoned, it was an obvious power relation in which the company's interests always carried greater weight.

At the end of the 1970s and beginning of the 1980s, the iron and steel industry was radically reorganized due to a structural crisis in this sector. As a result of closures and new technology, the total number of employees in the Swedish mining industry fell from about thirteen thousand in 1975 to about four thousand at the turn of the century.[37] Unemployment became a considerable problem as mining activities in Malmberget receded. In the 1980s, the Gällivare municipality tore down several multifamily houses in Malmberget, corresponding to five hundred apartments for lack of tenants.[38] The future looked bleak.

The Town and the Company at the Turn of the Twenty-First Century

For decades the future prospects in Malmberget had been most uncertain and generally painted in gloomy colors. In the autumn of 2003, yet another municipal comprehensive plan for Gällivare and Malmberget was adopted, albeit with very little political engagement, first because planning was not regarded as a high priority, and also, one might assert, because the very idea of planning is not well suited to situations of economic decline and shrinking populations.[39] At this moment, the world market price of iron ore suddenly rose dramatically, by 135 percent during the period 2003–2007, so that several of the ore bodies in Malmberget that had previously been abandoned immediately became profitable for mining again.[40]

The brand-new comprehensive plan was outdated in a few weeks and hope was rekindled in the local community as new work opportunities came in sight, although it was apparent that modern mining techniques required fewer people than before. The price increase furthermore implied that the two ore bodies stretching underneath central parts of Malmberget were to be mined more actively again, meaning that the buildings located above would have to be evacuated. The pace and extent of the evacuation was difficult to prognosticate, however, since at this time iron ore extraction took place at a higher speed and on a much larger scale than ever before. In parallel, LKAB carried out additional ore surveys that were to potentially affect even more parts of the town of Malmberget.

In spite of the rapid recovery of the mining industry, at this time Malmberget was in decay. The number of inhabitants was six thousand, one thousand of whom worked in the mine.[41] The town center contained empty blocks and young people moved away to pursue education and work (figure 2.2). Gällivare had decisively stepped forward as the more vibrant of the two towns, with about twice as many inhabitants; it also housed the municipal administration and the local

Figure 2.2 Malmberget's present town center was built in the 1960s. It replaced the old town center that had to be abandoned because of unstable ground. Today, many of the commercial premises are empty, and the high-rise building is decaying. Photo: Anna Storm, 2007.

museum. An 18-year-old girl described the Malmberget town center as a ghetto: "Only drunks and junkies live there...It is like a ghetto." Still, she thought there was something special and attractive with this ghetto in comparison to Gällivare: "It is kind of desolate, a bit hard and a bit cold, it is a bit remote. It is rather special. Gällivare is so, you know, ordinary, for everybody."[42] However, from an official point of view, the local government commissioner Tommy Nyström did not see any potential in the dilapidated Malmberget center. Instead he expressed a wish that it would disappear completely: "I hope they find valuable ore beneath every single house up there, so that we can demolish the whole town. Because if this doesn't happen...Malmberget will simply become a slum."[43]

What was going on in Malmberget was truly troubling. How was the municipality to deal with the maintenance of the indoor ice rink, the public baths, the schools, and the general infrastructure when it was impossible to predict which parts of the town—or perhaps the whole town—had to be abandoned? And if so, when would this happen? In the municipal planning documents after the dramatic iron ore price increase beginning in 2003, for example, it was noted that "a radical change of the comprehensive plan will be required in approximately three years' time," and even during these three years the plan was said to be in need of continuous revision depending on the results of LKAB's ongoing ore surveys.[44] Indeed, the ambiguous planning process of the 1950s and 1960s was echoed half a century later.

On the one hand, the municipal planning of the early twenty-first century was portrayed as a powerless reaction to "the problems that arise in relation to LKAB's development" on the one hand and, on the other, as the ambition to advance a municipal vision of the future, independent of the company.[45] This vision was based on the aspiration to reverse the negative population trend by offering people a high quality of life, while the question of employment and the town's dependency on one large enterprise was kept firmly in the background.[46] At the same time, however, the heavy reliance on LKAB was evident in such expressions as "the hand that feeds us" or "the mine is our bread and butter," used by municipal representatives.[47]

In general, there were strong expectations among municipal representatives and local inhabitants in Malmberget that LKAB would pay and take responsibility for activities other than mining and also replace the infrastructure and buildings that were affected by mining activities. For example, the local government commissioner Tommy Nyström expressed hope that LKAB would contribute financially to new public baths and an indoor ice rink when the existing ones eventually had to be closed in Malmberget.[48] These expectations were also based on the legal framework regulating mining activities in Sweden, primarily the Mine and Minerals Act. Since mining in some areas is designated as of national interest for mineral extraction, the law actually supports the compulsory evacuation of inhabited ground for the benefit of mining. A few conditions have to be fulfilled, among them that the ore body in question is estimated to be profitable to mine for a very long time, that the mining company is economically viable enough to bear all the costs connected with the forced evacuation, and that the municipality join in by changing its town planning regulations.[49] However, the verification and implementation of these conditions can be diffuse and subject to negotiation.

Besides, LKAB was still responsible for a number of activities in Malmberget, most of which had long been managed by the public sector. For instance, the company had its own water and sewage system in the company land area as well as the road networks and street lighting it owned, maintained, and financed.[50] LKAB also continued its involvement in the upper secondary school's mining program and every year the company guaranteed employment for ten students upon completion of their education.[51] For example, when most big companies in Sweden halted the construction and provision of housing for their employees at the end of the 1970s, for recruitment purposes, LKAB did not.[52] In the early twenty-first century LKAB's property company was responsible for 18 percent of the municipality's tenants. However, the building stock was maintained only as long as it was expected to be used for a further five years.[53]

In 2006, due to the upswing in the mining industry and the evacuation of some residential areas, Malmberget faced a housing shortage. LKAB had about eight hundred apartments rented and about five hundred people on the waiting list. Municipal and private property owners all reported long apartment queues.[54] Despite the lack of housing, the will to invest in new buildings was generally low on the part of both private and public property owners, who were all too aware of the recent demolition of five hundred apartments.[55] In addition, the municipal vision for the coming decades did not really involve investments in new housing, but instead stated that "mining spreads and this implies that the town will ultimately be divided," so that in the long term, the whole town might need to be closed down.[56] Hence, the threats and promises of mining activity were renewed.

The Company Land and Local Heritage Processes

What was the role of heritage in the midst of the critical changes characterizing Malmberget at the beginning of the twenty-first century? The narratives conveyed by local heritage institutions all showed a remarkable concordance. The municipal museum of Gällivare, the LKAB company museum, and two municipally organized heritage trails were dominated by the period prior to World War II. There were three main themes to be found: the culture of the Sami people, the hard life of the mining settlers at the turn of the twentieth century, and iron ore as a natural resource.[57] Alongside these articulated narratives, in 1987 the national heritage authorities designated the company land as of national interest for heritage conservation, which was later reasserted by the municipality in its conservation plan. But how did these narratives correspond to conceptions of meaning and value of the inhabitants in the transforming town of Malmberget?

In 2002, before the dramatic increase in the world market price for iron ore and the new future prospects in the local community, the County Administrative Board remarked in response to the ongoing comprehensive planning that the municipality had not accounted for its more modern heritage. The board stated that in times of recession or decline, culturally and historically valuable buildings hold symbolic importance for citizens' and the town's identity.[58] A local heritage association, too, was of the opinion that certain postwar buildings were worth preserving, and that the buildings should remain on their original sites and in "their proper historical context."[59] The municipality agreed that the existing conservation plan from 1985 was outdated and that a new plan should be established that took postwar buildings into account, but at the same time the municipality also emphasized that "all older buildings could not be preserved."[60]

With the changed prerequisites of 2003, when the price of iron ore drastically rose, the issue of building conservation became linked directly with the expected forced abandonment of certain residential areas and the possibility of relocating single or groups of buildings away from the unstable mountain. In the municipal planning the "historical built environment" was considered a potential resource to enhance the attractiveness of the town of Gällivare. This boost was to be achieved by moving key buildings from the company land in Malmberget to safer ground in Gällivare.[61] As an experiment, LKAB had tried moving a smaller number of modern homes from Malmberget to the "Mellanområdet" area between Malmberget and Gällivare—an experiment that was regarded as successful from an economic point of view.[62] However, LKAB contended that moving the older buildings from the company land and continuing to manage them in Gällivare was not feasible from the company's perspective. Instead, LKAB deemed the company land, as long as it remained intact, to be a resource for recruiting. The beautiful older villas were used to attract new personnel with special skills who might be hesitant to move that far north. Apart from recruiting, the control of the area was crucial for extracting the ore underneath, should this turn out to be profitable. If this happened, the company would not need to negotiate with other land owners before the necessary evacuation could begin.[63] Furthermore, the company land did not fall under municipal planning regulations, and as a consequence the municipality had quite limited opportunities to impose any measures regarding the company land being designated as heritage.

The obstacles to moving buildings on the company land to another area were not only technical and economic. Representatives of both LKAB and the town of Gällivare commented that moving a group of the older buildings elsewhere would also prove complicated because it would be difficult to recreate them in their entirety and in their natural settings. The birch-covered southern slopes around the company land in Malmberget had no obvious counterpart in Gällivare, and the buildings were not regarded as suitable for relocation as single entities since they were considered as a cohesive ensemble.[64] In practice, representatives of the municipality did seem to regard moving the older wooden villas to be utopian, both from a preservation point of view and from an economic and technical perspective.[65]

In the planning documents the company land was indeed proposed for conservation, but a municipal representative asserted that this would not make any major difference regarding the buildings' future.[66] The actual ambition was not to preserve the area, but rather to preserve a few pieces for the sake of memory: "We will never get the company land back, but fragments and single buildings can be preserved through relocation. The fragments can help us experience how it once was."[67] The somewhat wavering positions adopted by LKAB and Gällivare municipality can be understood as lip service paid to regional and national heritage authorities asking for modern conservation approaches, but it might also be a question of cautious examination. What was actually feasible? What could one think in a situation like this? There were certainly not many forerunners to learn from.

In addition, both LKAB and the town of Gällivare held the Swedish state primarily responsible for the protection of any heritage values connected with the changing town of Malmberget and with the company land in particular. The municipality contended that as the state owned LKAB, and because it was the state that had focused on the company land as a site of national interest for heritage conservation, responsibility for the historic environment within LKAB's territory should be placed at the national level.[68]

So what was considered important in Malmberget? To the local inhabitants, and to representatives of the municipality and the mining company, urgent themes of the past and present were connected less with the canonized stories found in the heritage institutions or in the older villas on the company land, and more with later changes like the mid-twentieth-century mining boom and the decades of decline.[69] The outstanding physical manifestation for this recent past was—without doubt—the Pit, the huge crater that had appeared in the midst of the town in the early 1970s and continued to affect its development thereafter. We will soon return to this decisive physical feature of Malmberget.

Kiruna on the Move

At the beginning of the twenty-first century, the population of Kiruna had grown to about twenty thousand, that is, slightly more than Gällivare and Malmberget together. In Kiruna we find both similarities and striking differences compared to the course of events affecting Malmberget. The 1970s saw growing interest in heritage perspectives, including academic perspectives on Kiruna's art and architecture, so that a body of canonized built heritage was soon to be distinguished.[70] This canon consisted of a number of key buildings, mainly originating in the early twentieth century and designed by famous architects or inhabited by famous men, such as the home of the LKAB manager Hjalmar Lundbohm, but also including the previously mentioned buildings from the 1960s, the multifamily block and the town hall. Furthermore, the canon highlighted the town plan ideas in themselves and the overall conception of Kiruna as a "model town." In the 1980s, the entire town had been designated as of national interest from a heritage perspective by the National Heritage Board, and a local conservation plan was adopted as well. Twenty years later, a general consensus about what was to be regarded heritage in Kiruna was shared by most of the town's inhabitants of all generations, reinforced by official statements from LKAB and the town of Kiruna.[71]

While the potential consequences of iron ore bodies stretching underneath the town center were a recurrent topic in Malmberget,

Kiruna faced fewer such concerns. In the 1970s, an area in Kiruna located close to the mine called "Ön," with housing, workshops, and storehouses, was emptied and dismantled due to predictions of unstable ground. This event was certainly not as dramatic or controversial as the evacuations and abandonment of the town center in Malmberget. However, in 2004 the fundamental prerequisites changed in Kiruna as well. At this time, as a direct consequence of the drastic increase of the world market price of iron ore, LKAB announced that mining activities would affect the ground underneath the town of Kiruna as well, prompting the municipality to state immediately, and loudly, that the entire town was to be moved in order to facilitate mining expansion.[72]

Unlike in Malmberget, where the protracted changes were debated mostly in a local context, the envisaged move of the whole town of Kiruna attracted spectacular attention nationally and internationally. Similar to Malmberget, the expected subsidence had actually been known for some time, not least by indigenous Sami people, and publicly forecasted by LKAB since the 1970s.[73] However, in Kiruna, too, it seems as if the obvious consequences of the predictions had been neglected by a majority of the inhabitants, so that the local reactions in 2004 expressed shock and frustration.

The undertaking to move the entire town involved a complex weave of so-called areas of national interest. No fewer than eleven such areas overlapped within the extensive town area, including mining, cultural heritage, and reindeer herding—each of which calls for different kinds of management.[74] Applying the work by heritage scholars Laurajane Smith and John R. Pendlebury, architectural heritage researcher Jennie Sjöholm has argued that the situation in Kiruna illustrates how an "authorized heritage discourse" is often challenged not primarily by subaltern discourses, questioning the selection of heritage, but, instead, that heritage is one elite discourse fighting other elite discourses to claim the validity of its perspectives. In Kiruna, the elite discourse competing with heritage perspectives is, of course, the mining industry; in spite of the clearly canonized body of heritage in the town, the mining discourse is unquestionably predominant.[75] The assistant town architect in Kiruna, Thomas Nylund, summarized the situation in a few sentences: "What is more important, buildings valuable for cultural heritage reasons or the extraction of iron ore? In this particular question the answer is given. That is why we have to move or replace parts of Kiruna."[76]

How could the answer be given? It was as given as it was in Malmberget, based on the legal framework for mineral exploitation and the conditions for the forced relocation of built environments.[77] However, the head of the Mining Inspectorate of Sweden, Jan-Olof Hedström, admitted that the number of different laws relevant to the case of Kiruna would make the process extremely complicated and turn out to be a full-scale test of whether the laws could actually work sufficiently when combined in such an intricate way.[78]

In spite of the predominant mining discourse setting the scene for local future prospects, to a large extent the planning to move Kiruna amounted to a debate about heritage. By law, LKAB was obligated to compensate for the damage and negative effects caused by its activities, implying, among other things, that built cultural heritage must be preserved as long as costs were "reasonable," a description that is certainly difficult to define in practice.[79] The canonized building stock and model town ideas were agreed upon by all parties; what was contested was how to best manage this designated heritage.

Two main groups could be discerned. On the one hand, LKAB and the municipal politicians agreed to single out a number of key buildings—around twenty at most—that were to be moved to the new town location; for the rest, the aim was to build the "model town, version two."[80] That is, both the company and local politicians emphasized the planning *idea* as the main heritage to carry on to the new Kiruna. On the other hand, the County Administrative Board, supported by local interest groups, argued for a larger number of buildings to be moved, in order to preserve as much of the town's present appearance as possible.[81] Jennie Sjöholm highlights how both parties thus wanted to recreate the "model town," but in different ways.[82] Most of Kiruna's residents did not express their specific opinion in any noticeable way, something the municipality interpreted as a sign that they did not care, while interest groups argued the opposite.[83]

LKAB and the municipality thus shared a general view on how to manage the relocation of the town. However, conflicts rose in connection to the practical planning process, and concerning the actual layout of the new town plan.[84] When the municipality was about to present its first considerations and suggestions, its work was suddenly interfered by an alternative plan, designed by a consultant commissioned by LKAB. The assistant town architect Nylund exclaimed: "Many people wonder who is actually planning? Is it the municipality or is it LKAB?"[85] Thereafter, two subsequent decisions were made on the new location of the town, the latest in 2011, which advocated a location east of the present town center. Step by step, infrastructure has been relocated, such as railroad and sewage and power lines, and the first residential areas in present-day Kiruna will disappear in 2014.[86] The areas that will successively be abandoned are to be transformed into a park—"the mining town park"—where artistic installations will offer viewpoints and spaces for contemplation and recreation until the area finally has to be fenced and closed off to any human presence.[87]

The Pit—Simultaneously "Feeding" and "Eating" the Town

The crater in the midst of Malmberget has been a prominent part of the town's narrative since the 1970s. At that time, an originally open cast mine started to grow due to explosions and cave-ins caused by underground mining activities. The crater was first called "the Captain Pit," because of its location above an ore body named the Captain ore body, and later just "the Pit," as no other mining cavity affected the town center so decisively (figure 2.3). At the beginning of the twenty-first century, the Pit had become about two hundred meters deep and covered an area of 21 hectares. Surrounding the Pit

Figure 2.3 "The Pit" is a crucial feature of Malmberget. Its crater has affected the town since the 1970s, dividing it in two halves and forcing the inhabitants to abandon homes and public buildings. Empty roads indicate where the former town center once was. Photo: Anders Andersson, about 1980. Courtesy of Gellivare Bildarkiv.

was a landscape of annual rings of fences, marking its continuous growth. The crater itself was almost invisible from a surface perspective, due to vegetation, so that all you could see when walking close by was young trees and roads that ended unexpectedly.[88]

There are many witnesses who emphasize memory places of their youth in connection to the Pit: "You have special feelings for the Pit because it is your childhood home." Such dreaming of days gone by was marked by both high sensitivity and lamentation:

> The garage, the laundry room, the boiler room, the kitchen, the lounge—all these smells are ingrained in my mind and the doors fling wide open as soon as the key turns in the lock... Others have become reconciled with what I experienced as a catastrophe for the town. They are born, live and love alongside the protective fence and are not interested in listening to an outsider's trips down memory lane.[89]

Young people instead talk about the Pit as if it were a living being they had to learn to relate to; an example being the expression "the Pit is coming," illustrating how the crater expands closer and closer to one's home.[90] Another significant experience deals with underground explosions in the mine, which take place every night at 12 o'clock. Eight-year-old children say that when they were younger, they were afraid of the noise and vibrations, but now they have become used to it. The Pit also bears mythical aspects, for example, illustrated by depictions of monsters explaining why there is so much rumbling in the Pit in the evenings.[91]

The Pit is also a place that is used actively by people in Malmberget, in spite of the fences and of the fact that access is strictly forbidden. The activities behind the fences range from quiet walks with the dog and searches for Christmas trees, to adventurous bike tours in the young forest and climbing on one of the inner walls of the Pit.[92] The Pit is dangerous, and therefore attractive, but certainly also frightening: "Yes, it has a bit of a Chernobyl feel to it, because it is kind of deserted and roads lead in the direction of the Pit and suddenly come to an abrupt end and houses just stand empty."[93]

The local government commissioner Tommy Nyström describes the Pit as the only thing that actually distinguishes Malmberget from other places and makes it special.[94] He underlines that the Pit is something really worth seeing, something he shows every important visitor, preferably from a helicopter, but also by going down into it by car. The invisibility of the Pit from the surface, which calls for helicopters and cars, is indeed striking. In the late 1980s, at a cost of several million Swedish crowns an old hoisting tower was restored to create a viewing tower overlooking the Pit. However, without any prior notice, LKAB suddenly fenced off the hoisting tower for use.[95] The reason was probably unstable ground, but the lack of communication did cause frustration in the local community, which was later revived when another old hoisting tower was suddenly burnt down in a training exercise for LKAB's own fire brigade.[96]

For LKAB, the Pit causes many problems. It emits dust, especially when cave-ins occur or waste rock is transported. Since the dust covers nearby windows and laundry, the company has been under pressure to reimburse house owners for the inconvenience.[97] The company has also worked to diminish the amount of dust, for example, by irrigating roads during transports and by new planting and wind protection nets.[98] However, in 2005 the company admitted that the efforts so far did "not delimit the dust proliferation that appears from time to time and which originates in part in the moraine cover of the pit edges."[99]

Furthermore, in spite of the testimonies by children asserting that they had become used to it, the vibrations in the ground caused by underground explosions at night also bring discomfort and worries to Malmberget's residents. LKAB responds by performing regular measurements in the homes closest to the explosions, but the municipality comments that the seismic activity will most probably increase in the future.[100] Moreover, the security issues with fences, and the prevention of border transgressions, place great responsibility on LKAB, as do negotiations about how to compensate those who have to move and leave their homes because of the Pit's expansion. The economic value of the houses located just outside the fenced danger zone around the Pit can be assumed to have dropped substantially, although at the same time they are not immediately threatened and therefore do not obviously fall within the framework of LKAB's financial responsibility stipulated by law.[101]

Since the area to be evacuated first due to mining at the beginning of the twenty-first century is not part of the company land area, LKAB will have to buy all the private properties before evacuating and dismantling the buildings. In communication with the property owners, primarily individuals owning a single family home, LKAB stressed the longstanding practice of radical change in the town, in order to historicize, and probably relativize, the drama of the situation. The company stated: "Malmberget has adapted to the mining activities for hundreds [*sic*] of years. The mine is our viability. Without it, Malmberget would hardly have existed. Homes and buildings have

Figure 2.4 As an experiment, a number of modern villas were moved from Malmberget to safer ground in Gällivare. The method proved technically difficult and too expensive compared to new construction. In this picture a youth choir leads the transport while performing a specially composed song. Photo: Marie Ridderström, NSD, 2007.

often been moved. And now, we have to do it again."[102] The plan was described as to move or tear down all the houses and then prepare the area for greenery, fencing it off and defining it as a risk zone for the foreseeable future.[103] The local government commissioner Nyström commented that it was actually a local tradition to say, "We have always moved," which he interpreted as a chastened attitude and one of the reasons for the quite limited public protests associated with the compulsory evacuation.[104]

Most of the families that had to evacuate their homes were offered the option of relocating the whole house. As previously mentioned, LKAB moved a few single family houses as a trial and evaluated the result as good.[105] The company described the method as simply lifting the house, putting it on a trailer, transporting it to a new plot where it was to be connected to water, sewage, electricity, and telephone (figure 2.4). The whole process was estimated to take less than a month. LKAB paid for the move, and, in the meantime, the family was offered a free substitute apartment and free storage of household goods.[106] A few years later, this method was abandoned for the most part, due to technical difficulties and low economic viability compared to new construction.[107]

In 2000, LKAB had started to fill the Pit because it needed a waste rock dump, and in order to prevent cave-ins—along with the more formal reason that "after-treatment" was prescribed by the legislation, obligating it to restore the landscape.[108] At a pace of eight truckloads an hour, day and night, the filling was estimated to take ten years.[109] However, due to the unstable ground, the company did not entertain the prospect of recreating land that could be built on, but only an area for walking and biking.[110] Later, the company stated that the Pit might in fact never be filled completely because of the continuous mining activity underground.[111] Instead, it may just shrink in one end, and expand in the other.

What does the Pit signify in Malmberget? Most fundamentally, when the Pit grows the mine is in production, which means work and survival for the people of Malmberget. At the same time it means danger, dirt, and forced relocation. It also illustrates the local community's heavy reliance upon the mining company in painful clarity. It is talked about as a scary but somehow familiar living being, and as the only thing that distinguishes the town from other places. In relation to the past, the Pit represents the golden age of mining and the prosperous welfare society of the 1950s and 1960s, as well as the former town center with its key public buildings and the lost places of childhood. The local attitude has nevertheless been described as "relaxed" in relation to the past and the future. While there is sorrow in the forced abandonment, there is also a "utopian energy" that emerges from this incessant movement.[112]

Furthermore, the Pit mirrors and defines the present declining town center. The Pit and the town center are both regarded as desolate, hard and cold, although also special and attractive because of these very characteristics. The uncertain future prospects of the town are physically taking shape in the Pit, with the omnipresent questions as to which part of the town will be the Pit's next "bite," and whether the Pit and the town will ultimately die together. A journalist formulated the situation as follows: "The sad thing is that Malmberget will probably die whatever happens. If all the ore is extracted, Malmberget will disappear, and if mining stops altogether there will not be any jobs and everyone will have to move."[113] As of 2013, the town of Gällivare is planning for Malmberget to disappear completely within the next twenty years, and relocate the inhabitants to Gällivare.[114]

Concluding Remarks

Spanning little more than a century, the town of Malmberget established and grew quickly, prospered and declined. It was a place for fortune hunters and a workers' community, and during the postwar decades an expression of Swedish welfare more than one thousand kilometers north of the capital of Stockholm. In the background was always "the hand that feeds us," the mining company LKAB, and in the periphery were the displaced indigenous Sami people. The relation to Kiruna was mostly a contrasting affair, with Malmberget the chastened, less spectacular cousin, while the relationship to Gällivare was initially marked by pride and distinction in the mining town and later by Malmberget as the "expendable community," while Gällivare stepped forward as the more vibrant of the two.

Malmberget represents an unsettled post-industrial scar. It is moving; underground, above the ground, with its people, and in its significance. The scar is epitomized by the Pit, the vast crater with organic qualities. Through its continuous change the Pit is never as it was, so different generations have met the Pit in different shapes. The Pit is both presence and absence. It is something obviously present in itself, a monster rumbling in the night and a territory of forbidden adventures and, at the same time, something absent, an emptiness where the former town center of the "good old days" used to be. The Pit illustrates how the affection that Malmberget residents might express toward material structures is always conditioned.

The reason for Malmberget's existence was a conscious choice to mine iron ore in the mountain to extract economic profit, and this logic has prevailed from the beginning to the present day. Accordingly, in a caption to an aerial photo of the Pit in a 2006 municipal planning document it was stated: "The pit in Malmberget is a historic imprint showing that the Swedish economy has been supplied with many billions of Swedish crowns," indicating that most of the money certainly did not stay in the local community.[115] The Pit is not articulated as important for collective memory, but it is indeed used as a memory trigger by individuals in the town. Its potential is an enhancement of shared memories explicitly connected to the landscape scar in the midst of the town of Malmberget.

Chapter 3
Distance of Fear

The closer to a hazardous industrial enterprise you live, the less afraid you are. Fear increases with geographical distance, so that from afar monstrous and even mythical characteristics are sometimes ascribed to an industrial activity. I posit that this paradoxical rule holds at least as long as no serious accidents occur at the industrial enterprise in question. This distance of fear has shaped some of the conflicts around commercial nuclear power, creating a "we" and a "them," representing fundamentally different worldviews between those living close to a nuclear power plant and those further away. The consequences of geographical distance are even more extreme when national borders are crossed, defining the understanding of a particular nuclear power plant as domestic or foreign, along with an assumed level of technological advancement and safety. Unsurprisingly, domestic plants are usually viewed in a more friendly light than foreign ones.

When it comes to the by-products of nuclear power, the distance of fear becomes more complicated. The immediate benefit of energy production is no longer present and the future prospects of contaminated ground where the power plant once stood, or the establishment of storage facilities for radioactive waste, evoke a wider range of reactions. The physical and mental landscapes scarred by radioactivity, by political conflict, and by fear are intertwined with preconceptions of the possibility to clean up, to control, and to exert responsibility. One response from the nuclear sector to the task of dismantling the first generation of commercial nuclear reactors has hitherto been to aim to restore the site to its previous condition. That is, the goal is to erase all traces of the plant, of the workplace, of the target of nuclear fear and instead reinstate a "natural" landscape. The physical scar is therefore meant to disappear, or at least become invisible. The mental scar is indeed ambiguous and borne by groups with quite disparate views on nuclear power.

This chapter will explore the history of the Barsebäck nuclear power plant, located on the southwest coast of Sweden, and the trust, pride, and fear connected with it at various geographical distances. Barsebäck was planned and built in the 1960s and 1970s, initially a period of strong optimism in the future and belief in technological progress. Soon, however, political and public winds shifted, and Barsebäck came to symbolize the threats of nuclear power in both Sweden and neighboring Denmark. The story of Barsebäck is played out in between connotations of, on the one hand, a wonderful Christmas gift to the region, an appreciated workplace, and part of a major national scheme of nuclear power and, on the other hand, a hated silhouette from the viewpoint of the Danes and a key symbol for disputes about energy politics in Sweden.

A Christmas Gift

Two days before Christmas Eve 1965, representatives for the town of Löddeköpinge in southern Sweden were called to a meeting at the County Administrative Board to receive truly surprising news. After three years of secret preparations, the energy company Sydkraft was about to build a nuclear power plant in the municipality. The information came as a "bombshell" since nobody had known what was going on behind the scenes.[1] The municipal representatives expressed much enthusiasm and claimed to "stand affirmative in every respect" toward this "terrific Christmas gift."[2]

In the 1960s, the establishment of the Barsebäck nuclear power plant was explained by the increasing energy consumption in Sweden—of which "no limit can yet be discerned"—in combination with the expected scarcity of other types of energy such as hydropower, oil, and coal.[3] The nuclear power plant was furthermore seen as just one of several similar plants to be built in a "string of pearls" along the southwest coast of Sweden, preferably close to population centers where consumption was high.[4] The proximity to consumers was also crucial since there were plans to provide district heating from the nuclear plants.[5]

This part of Sweden is a relatively densely populated, flat agricultural landscape scattered with smaller settlements and several bigger cities and towns. The area where the nuclear plant was to be built was used mainly for pasture and regarded as of great natural beauty because of its open view over the Sound—the inland sea between Sweden and

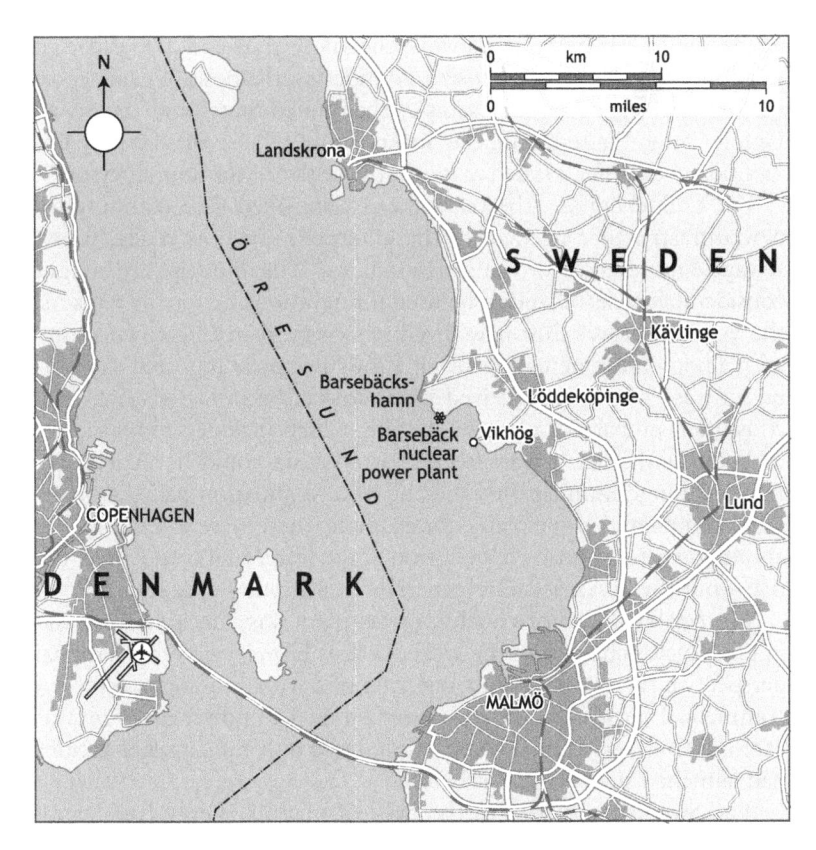

Map 3.1 Barsebäck nuclear power plant is located on a cape in Öresund, close to the cities of Malmö, Lund, and Landskrona. Twenty kilometers across the Sound is the Danish capital of Copenhagen. Map: Stig Söderlind, 2013.

Denmark. Two small fishing villages—Barsebäckshamn and Vikhög—were the closest settlements. The main town, Löddeköpinge, is located about 7 kilometers from the site chosen for the plant, while the cities of Malmö, Landskrona, and Lund are all within 30 kilometers. The Danish capital of Copenhagen is 20 kilometers (map 3.1) away across the Sound, decisively closer than the Swedish capital of Stockholm, located 590 kilometers up-country.

All land within 500 meters of the proposed plant was bought by Sydkraft, and a safety zone of a couple of kilometers was established in which no permanent housing was to be built. However, the safety zone was seen as ideal for holiday cottages and there were plans to develop the area for pleasant country paths and similar facilities in the near future.[6]

According to Sydkraft, after a while the safety zone could be expected to be significantly reduced in size, considering the positive experiences with nuclear plants in, for example, the United States and the Soviet Union.[7] Sune Wetterlundh, the Sydkraft CEO, believed that they had "found an ideal site, located as it is in the center of the Sound region."[8] For Löddeköpinge, the years ahead were filled with planning for new infrastructure needed by the enterprise, such as roads, water, sewage, and reserves for transmission lines.[9] The municipality bought considerable areas of land to be used for new housing for the workers; the estimation was 200 new dwellings every year, in a town center with a population of just 650 as of 1960. The press reported that this boom was "of course, received with satisfaction in the district."[10] A consultant prepared a plan for the site, which included four reactors to begin with, but it was noted that even six could be expected.[11] Löddeköpinge municipality and the local population had only marginal comments on the plan—for example, there were some concerns about potential radioactive pollution of the water that could affect the fish—but these worries were generally dismissed as exaggerated.[12]

The establishment of a nuclear power plant was a decisive event and implied major changes in the local context. However, from a national perspective the plant was just one part of a wide-ranging scheme for commercial nuclear power. The Swedish nuclear power program was extensive in international comparison, since only much larger nations had launched such programs before.[13]

The Swedish activities in the field of nuclear energy had begun directly after World War II and targeted both military and civilian uses. Research was carried out by a state-led company, and in 1954 a small research reactor went critical for the first time. A long-term program was adopted by the parliament in the next year, with the goal of creating a domestic nuclear fuel cycle, including extraction of uranium, the construction of nuclear reactors, and technology for reprocessing spent fuel. The larger aims were to diminish dependency on foreign energy supplies and to be able to construct nuclear weapons. The military aspect was controversial, however, and not discussed publicly. What became known as the "Swedish Path" entailed technical choices to focus on heavy-water reactors with domestic natural uranium in order to meet the overall goals of the program.[14] In the mid-1950s, it was envisaged that six smaller reactors for district heating purposes and one full-scale nuclear power plant would be built within the coming decade. STS scholar Jonas Anshelm has characterized this period as a visionary phase during which the nuclear "demon" was to be tamed by the scientific "wizards." Any risks were expected to be handled by technology and through a learning process, similar to how society had adapted to electricity.[15]

Soon enough, the enthusiasm was tempered by frustrating delays and increasing costs and the plans were scaled down. During this "practical realization phase" a small heavy-water reactor began operating in Ågesta in 1963, providing the Stockholm suburb of Farsta with district heating, and in 1970 a full-scale reactor was finished in Marviken.[16] Ultimately, due to failure to meet safety requirements, the Marviken reactor was never commissioned. At this time, private actors had already chosen to invest in light-water reactors instead since, according to leading power companies in the United States, these had proved superior from a commercial point of view.[17] Finally, the government-led investors also abandoned the "Swedish Path." Switching over to light-water technology also meant discarding any military ambitions. In spite of the Marviken fiasco, in the early 1970s the nuclear power industry had bright prospects for the future. In 1971, six full-scale light-water reactors were ordered, mainly from the Swedish company ASEA, and the branch planned to build 24 reactors in the coming two decades. Among those ordered was the first reactor in Löddeköpinge, Barsebäck 1.[18]

At this time, the issue of risk connected to nuclear power had begun to enter the public agenda, but trust in the technology and general enthusiasm dominated. In connection with the Barsebäck plant, a director at Sydkraft assured that rigorous safety regulations in fact made nuclear power plants the *least* risky of all industrial sectors, that the radioactive waste was "easy to control," and that the spent fuel was to be transported "at sea in a way that ensures complete safety."[19] It is also important to note that the most highly radioactive by-product, spent fuel, was not regarded as waste until the mid-1970s, but rather as a valuable resource for the next generation of reactors.[20] In addition, the estimations of the time needed to store radioactive waste changed dramatically during the second half of the twentieth century. In the 1950s, it was expected that radioactive waste had to be stored isolated from humans for less than a hundred years. During the 1960s the estimations changed first into thousands of years and later, into hundreds of thousands of years.[21]

In several countries, nuclear power plants were viewed as objects of architecture; for example, in Great Britain and France well-known architects were hired to give the new industrial establishments a qualitative architectural form.[22] In Sweden, however, this was not the case. Architects were indeed commissioned to design canteens and exhibition pavilions in connection with nuclear power plants, but the plant itself was seen as a strictly technical structure.[23] Art historian Fredrik

Krohn Andersson argues that, for contemporary actors, the significant visual context for Barsebäck was the landscape, in terms of both location and scale.[24] Barsebäck nuclear power plant was to become large and visually dominating, located along the coastline, and this circumstance was thoroughly examined by the planners, who emphasized the visual contrast as something positive.[25] Representatives for heritage authorities, too, expressed similar affirmative opinions about the new industrial construction. The Swedish local heritage federation carried out a rather detailed account of the cultural history and natural environmental characteristics of the area, and a number of archaeological sites were excavated before construction began.[26]

The most important ancient monument was a passage grave from the Stone Age—Gillhög—located on a small hill a few kilometers from the plant site. In order to protect the grave and its context, the National Heritage Board, among others, stipulated that the plant had to have a sophisticated form that would harmonize with the grave hill and connect to the cultural landscape, and also that the transmission lines should be kept as far away as possible, on poles no higher than the hill.[27] Thus, there was no serious conflict between the existing landscape and the proposed new industrial infrastructure, but merely a question of correspondence.[28]

Early in the construction process, a specially dedicated exhibition pavilion was built next to the construction site since the plant, this "energy source of the future," was expected to become a big tourist attraction immediately.[29] The exhibition pavilion was to house information about nuclear power technology and to show findings from the archaeological excavations.[30] The initial incorporation of the nuclear power plant under construction into the landscape was thus a smooth story for the most part, in terms of local attitudes, heritage, and visual appearance.

Shaping Local and Regional Trust

The power plant construction began in 1970 and employed about 250 men, a number that would increase to 700 later on when the reactor vessels were put in position.[31] The location of the plant implied short distances to electricity consumers in southern Sweden, and the neighboring country of Denmark was also included in the market envisioned for the Barsebäck plant. In an agreement between Sydkraft and the Danish energy company Elkraft, Barsebäck was to provide electricity to Denmark until the Danish nuclear power program was up and running; after that, a mutual exchange of nuclear electricity was foreseen.[32]

DISTANCE OF FEAR 53

The regional cooperation in electricity exchange between Sydkraft and nearby Denmark had been established long ago. Sydkraft was founded in 1905, owned mainly by five cities in southern Sweden, with its activities based on hydropower.[33] Early on, there were plans to connect the Sydkraft grid to a grid on the Danish side, so a submarine cable was laid in the Sound in 1915.[34] When there was an abundance of water in the southern Swedish rivers, the generated electricity surplus was sold to Denmark, and in times of water shortage, Danish thermal power plants became a backup for Sydkraft. As Sydkraft grew, two additional submarine cables were laid in the late 1920s and another three in the 1950s, financed mainly by the Danes.[35]

In 1957, a first nuclear research reactor in Denmark went critical at the state-owned nuclear research station Risø. During the subsequent decades, steps were taken to introduce commercial nuclear power.[36] Just as in Sweden, all the Danish political parties were positive about

Figure 3.1 As a token of the close and friendly relations between Denmark and Sweden in the energy business, a Danish girl orchestra plays at the 1977 topping out party for reactor number two at Barsebäck. Courtesy of E.ON Kärnkraft Sverige AB.

this new source of energy, as was the press.[37] To extend the already established cooperation in the electricity market between Sydkraft and Danish companies into the field of nuclear power was therefore nothing surprising (figure 3.1), and yet another transmission cable between the two countries was laid in the early 1970s.[38]

At the local level, Löddeköpinge cooperated with Sydkraft on issues such as a new water purification plant and street lighting, and the population of the local community increased considerably.[39] Within a fenced area, two boiling water reactors in two buildings the height of a 20-storey house and two chimneys rising 120 meters above the ground gradually took shape on the cape between the two small fishing villages (figure 3.2). The facades were covered with gray aluminum plates, broken up by a pattern of black vertical stripes. According to contemporary press reports, the idea was that the stripes would make the plant visually merge with the horizon and the sky. Some also claimed that the graphic pattern illustrated the fuel elements and the core.[40] Outside the fence were, among other things, a

Figure 3.2 Initially, the Barsebäck nuclear power plant was planned to consist of four, or even six, reactors. The closest settlements are Barsebäckshamn (in the foreground) and Vikhög (discernable in the background). Photo: Richard Conricus, 1989. Courtesy of Bilder i Syd.

distribution plant, a reserve power aggregate, and buildings for temporary residents.

In 1974, the town of Löddeköpinge merged into the larger municipality of Kävlinge. While the biggest employers in this new administrative unit were connected to farming (a slaughterhouse and a tannery), the nuclear power plant nevertheless established itself as an important employer in the area, with about 350 on the payroll.[41] The first Barsebäck reactor went critical in 1975 and in the fall of this year, Barsebäck was already providing electricity to Denmark.[42]

Almost from the beginning, Sydkraft and the local community established a close and trustful relationship.[43] Apart from the aforementioned infrastructure investments, the nuclear power plant also required a full-time manned fire station in the municipality and, like many other companies, Sydkraft began to sponsor local sports clubs, invite school classes to visit the plant, and inform about its production through advertising in newspapers and leaflets sent out regularly to all households.[44] One very popular element of the communication between the company and the local community later on was an annual photo calendar featuring romantic depictions of the plant along with nature motifs.[45]

Nevertheless, living in the vicinity of a nuclear power plant also entailed some specific everyday peculiarities. A fine-meshed alarm system was implemented to respond to emergencies. It was first based on AXE (telephones) and later on RDS (radio data system) technology; more than thirty thousand RDS receivers were distributed within the so-called emergency zone around the plant.[46] Another safety measure for the local area was the distribution of iodine pills by employers and by the pharmacy.[47] In retrospect, these material reminders of the particular risks of nuclear power were denoted as mere routine by municipal authorities, and regular opinion surveys showed that a vast majority of those living in the vicinity had high confidence in the Barsebäck plant.[48] It is certainly possible to interpret these assurances of trust as strategies to deny or neglect the risks. However, the conscious efforts by Sydkraft to win over the local population seem to have been rather successful in building a confident relationship with the local community.[49]

This picture is also strengthened by the work carried out by what was called the local security committee, initiated by national authorities as a means of exercising some public control over the plant. The long-standing local government commissioner Roland Palmqvist, for example, characterized both the relations to Sydkraft and the public control as "very good...thanks to the local security commission."[50]

On one rare occasion, the committee was criticized—from within. The vice chairman of the committee asserted that it was in fact a "taboo" to question the safety of Swedish nuclear power plants in any way, and implicitly accused the committee of hiding inaccuracies and other problems at Barsebäck.[51] His views were not met with any sympathy, either by the rest of the committee, or by the plant management or national nuclear authorities, and he was later dismissed from the commission.[52]

The community interaction that took place because of the existence of the plant was characterized by both normality and a special identity. The nuclear power plant was in many ways just like any industrial enterprise in the area, but at the same time it brought special features that affected private homes as well as municipal politics and administration.

Antinuclear Winds

The year 1972 was a breaking point in Swedish nuclear debate. Up to that time energy politics was generally regarded as a domain for technical experts. After this year, scientists began to openly express diverging opinions, expert knowledge was questioned, and nuclear power became a political and moral issue of public debate.[53] An antinuclear movement took shape, inspired by similar activities in Germany, France, and the United States, and spurred by a large international conference on the environment held in Stockholm the same year. Thus, the construction of the Barsebäck nuclear power plant coincided with a fundamental change in the societal understanding of nuclear power. A radio play about an imaginary accident at Barsebäck, broadcast in 1973—even before Barsebäck's first reactor began operating—found an audience that, according to some sources, began to panic because they believed the accident was real (reminiscent of the Orson Welles broadcast of "War of the Worlds" in 1938).[54] According to a Sydkraft chronicle, however, people rather became "somewhat worried" and the "panic" took place only in newspaper reporting.[55]

The antinuclear standpoint also became influential in the second largest political party at the time, the Center Party.[56] Leading members of the Center Party were influenced by the respected physicist and Nobel Prize winner of 1970, Hannes Alfvén. Alfvén had been engaged in the national nuclear program from the start but later began to criticize nuclear technology. When the Left Party also declared an antinuclear position, the old left-right dimension in Swedish politics was challenged, and in 1973 the oil crisis moved energy to the very center of the political agenda.[57] In 1976, the Social Democratic Party was voted out of power for the first time in 44 years. A center-right coalition won the close elections, and the tipping point was likely the antinuclear standpoint of Center Party leader Torbjörn Fälldin. The nuclear issue caused much conflict within the new government, since the other parties in the coalition did not share the Center Party's antinuclear position. In 1977, a second reactor began operating at Barsebäck and simultaneously a Nuclear Stipulation Act was approved, which required nuclear plant operators to prove, before loading new fuel rods, that they could manage radioactive waste in an absolutely safe way. The implications of this act launched prolonged interpretative conflicts, and in 1978 the government was forced to split. Efforts to find a political compromise were interrupted in 1979 by the Three Mile Island incident in the United States or, as it is often called in Sweden, the Harrisburg accident.

The incident was given extensive attention in the press; compared to the US press, for example, Swedish reporting was much more pessimistic.[58] A reporter interviewing two young women working with information for the public at the Barsebäck plant at the time presupposed that their work was "therapeutic" since the nuclear issue caused such passionate opinions—for or against.[59] However, on the contrary, one of the women replied that many people were in fact indifferent to nuclear power. Although she had been asked a lot of questions after the Three Mile Island incident, few visitors were scared but instead "open to evident data."[60] One concrete consequence of the Three Mile Island incident was that a special filter construction was erected at Barsebäck, the so-called Filtra. It was a cylindrical building filled with pebbles that was to work as a strainer for radioactive particles in case radioactive materials were released.[61]

On the national level, the Three Mile Island incident became the last straw in a long fight about holding a referendum about the future of nuclear power in Sweden. The event in the United States finally convinced the Social Democratic Party leader Olof Palme to favor a referendum, probably in order to avoid a nuclear debacle in the upcoming parliamentary election. The antinuclear parties formed one alternative, while the pronuclear parties formed two alternatives with slightly different formulations. Technology historian Arne Kaijser asserts that this was for tactical reasons, that is, the intention was to offer an alternative that appeared as a middle way and thus attract more voters.[62] The three alternatives all formally favored a phasing-out of nuclear power in Sweden, although on different conditions, since that was the only politically possible standpoint at the time.[63]

The referendum was to take place in the spring of 1980 and the campaigns were intense and engaged large parts of the population.[64] At Barsebäck, employees remember tight schedules—"three-four evenings a week"—of presenting and debating in study groups and hearings in connection with the referendum.[65] The basic message they brought with them was that knowledge was the key to accepting nuclear power and that "ignorance [was] behind most of it [the antinuclear attitude]," which was completely in line with the perspective advocated by most nuclear professionals.[66] After a nail-biting election night, the "middle way" alternative won with 39.1 percent, the antinuclear alternative got 38.7 percent, and the more openly pro-nuclear alternative got 18.9 percent. According to Sydkraft, people who lived closer to Barsebäck generally tended to vote in favor of nuclear power. The company commented that in "the cities of Malmö, Lund and Landskrona people were less afraid than in Norrland [a northern part of the country]."[67] A distance of fear was articulated in the referendum.

A few months later the Parliament decided that all nuclear power plants should be phased out by 2010 and that alternative energy sources and energy efficiency should be encouraged and enhanced to fill the vacancy. The goal of 2010 was based on the assumption that the last of the twelve reactors that had been ordered at this time would be commissioned in 1985, with an expected operating life of twenty-five years. In the long-term perspective, this decision implied major changes. However, in the short term, it meant more or less back to business for a sector that had been heavily disturbed by the period of political turbulence.[68] Nevertheless, after the referendum, the plans for Barsebäck 3 and 4 had to be abandoned.[69]

A Danish View over the Sound

Antinuclear opinions entered the political agenda in Denmark as well. Following the foundation of Organisationen til Oplysning om Atomkraft (the Organization for Information about Nuclear Power), OOA, in 1974, the issue was definitely at the center of attention in the Danish public discourse. The first action taken by the OOA was to claim a three-year moratorium on efforts to establish Danish nuclear power, justified by what the organization termed to be a need for further studies before making any decisions. Just two months after the foundation of OOA, twenty local sections of the organization had been set up all over the country, and the OOA became one of the predominant actors in the struggle over nuclear power in Denmark.[70]

During the first years of public debate in Denmark, Barsebäck was not the most frequently occurring theme. Instead, there were more general topics of discussion concerning the possible Danish commercial reactors, the consequences of radiation, nuclear bombs, and the transport of radioactive materials.[71] In 1977 and onward, however, Barsebäck became a major focus for the argumentation of the OOA, fueling discussions about possible locations of nuclear power plants in Denmark, about the Swedish transport of spent fuel through the Sound on its way to reprocessing in France, and about the probability and consequences of a major accident at Barsebäck.[72] For example, the OOA provided detailed calculations about possible contamination and the subsequent forced evacuation of large numbers of people from the greater Copenhagen area should there be an accident at Barsebäck, and later also presented an alternative energy plan for the country.[73] The two dominant themes of the debate over the years were the issues of radioactive waste and the safety and risks connected with nuclear power plants.[74]

OOA was certainly not unchallenged. The most prominent advocates of nuclear power in Denmark were the representatives of the electricity industry and the researchers at Risø. A separate organization, claiming to represent popular opinion in favor of nuclear power, was also founded: Reel Energi Oplysning, REO (Actual/Real Energy Information). Both the advocates and the opposition mainly used a language marked by rationality, technology, and natural science, although both sides were also prone to more emotionally loaded rhetoric.[75] For example, representatives of the OOA referred to leading nuclear advocates as "the nuclear priesthood," while at Risø the employees characterized fearful attitudes toward radiation as "hysteria."[76]

OOA has been described as a grass-roots movement, explicitly dissociated from political parties.[77] The activists wrote polemical articles, published a magazine and numerous brochures, and initiated demonstrations, marches, concerts, and theater plays.[78] Its most well-known symbol was a yellow and red badge with a smiling sun and the text "Atomkraft? Nej tak" ("Nuclear power? No thanks"). The symbol was designed by the young OOA activist Anne Lund in 1975, translated into 42 different languages, and over the following years fifteen to twenty million copies were printed.[79] In retrospect, Lund explained that it was a very consciously chosen symbol, not evoking radical left-wing aesthetics with clenched fists and red flags, but representing respectful dialog, kindness and politeness, "like a nice lady would like to wear on a nice trench coat."[80] The overall combination of technical competence and emotional activism was the philosophy of the organization.[81]

The OOA was financed mainly by volunteer work, receiving economic support from individuals as well as income from sales of the "sun badge."[82] In 1980, the organization counted 130 local sections in Denmark, with some of the demonstrations attracting up to twenty thousand participants, and on several occasions hundreds of thousands of signatures protesting Barsebäck were collected and handed over to the Swedish government.[83] The majority of OOA members were young, and the work they put into the struggle against nuclear power turned into a lifestyle that influenced them in a fundamental way, not least as a kind of informal educational institution.[84]

Barsebäck certainly was a key symbol for the OOA.[85] One of the most frequent slogans was "Hva ska väck? Barsebäck. Hva ska ind? Sol og vind" ("What should go away? Barsebäck. What should come? Sun and wind") (figure 3.3).[86] In the visual language, OOA offered future scenarios, for example, an sunny and verdant aerial view over

Figure 3.3 Danish banners were common in the yearly antinuclear marches from Lund to Barsebäck in the late 1970s and early 1980s. Some demonstrations attracted about thirty thousand participants. Photo: Torbjörn Carlson, 1980. Courtesy of Bilder i Syd.

the Sound area between Denmark and Sweden, where the site of Barsebäck was turned into a "memorial park," contrasted by a dark, heavily industrialized counterpart.[87] The very form of the Barsebäck plant was used as well, for example, on banners that showed the characteristic vertical stripes of the Barsebäck reactor buildings as the teeth of a skull.[88]

That there was actual fear among many of the OOA activists is clear, illustrated, for example, by the account of a big march to Barsebäck in 1977 written by thirteen-year-old Lui: "We came closer and closer to the frightening monument of fear of a nuclear power plant...suddenly we stood in front of this abominable death trap, which the Swedish government put up without showing any regard to the inhabitants of the area."[89] Paradoxically enough, the contact between Danish and Swedish antinuclear activists seems to have been relatively limited, in spite of joint demonstrations and marches. A leading OOA activist, Jørgen Steen Nielsen, says that the lack of close cooperation was probably due to cultural differences.[90] At the same time Swedish authorities, individuals, and publications were often cited in the Danish public debate, by both advocates and opponents of nuclear power.[91] Among the better-known names of antinuclear spokespersons used by the OOA we find, for example, the Swedish writer and comic actor Tage Danielsson performing a monologue about a "probability calculus" for the Harrisburg accident/Three Mile Island incident, and the physicist Hannes Alfvén unctuously pleading against nuclear power.[92]

Chernobyl and a "Premature" Shutdown

In 1982, the Social Democratic Party was back in office. However, any hope it harbored that the hot topic of nuclear power could be cast aside from the national political discourse was disappointed. Instead, the Social Democrats experienced a harsh internal fight between antinuclear groups within the party on the one hand, and representatives for the trade unions on the other. The latter were worried that phasing out nuclear power would lead to increased electricity prices, which would severely affect basic manufacturing and, in consequence, harm employment in this key sector.[93] When a government commission explored the possibilities to advance the timetable for phasing out nuclear power, this led to intense lobbying by the trade unions and the power industry against what they termed a "premature" shutdown.[94]

The explosion at the Chernobyl nuclear power plant in 1986 shocked nuclear experts and the public alike. The accident was first detected in the West by Swedish nuclear engineers; in fact, the Swedish city of Gävle, 1,500 kilometers from Chernobyl, received more contamination than Kiev, located just 130 kilometers from the accident. Leading Swedish politicians demanded a total nuclear phase-out as soon as possible, and an immediate shutdown of Barsebäck, while several environmental organizations demanded the immediate closure of all nuclear power plants.

The government responded by initiating an investigation on risk assessment. To some extent, nuclear experts succeeded in convincing both the public and the politicians that there was no real connection between Swedish nuclear power plants and the Chernobyl plant.[95] The differences between Soviet and Swedish reactors were brought to the fore and emphasized by nuclear advocates. The logic was that it was meaningless to phase out Swedish nuclear power when there were much more dangerous reactors located just on the other side of the Baltic Sea, among them at the plants of Sosnovjy Bor, Ignalina, and Greifswald. The pronuclear conclusion was to spend money where it could be of greatest value, that is, in Eastern Europe.[96] Sweden also became quite involved in safety upgrades, especially at the Ignalina nuclear power plant, but this work did not influence the national debate in any fundamental way.

In 1985, after more than a decade of intense debate, the Danish Parliament decided to exclude nuclear power from its comprehensive energy planning. In 1986, in response to the Chernobyl accident, the parliament furthermore instructed the Danish government to contact the Swedish government immediately with a request to close down Barsebäck.[97] In 1988, the Swedish Parliament, too, decided to close down two of the country's existing twelve reactors by 1995 and 1996. There were several concurrent reasons why the two Barsebäck reactors were chosen to be the ones closed. First, they were relatively small in effect, and closing down these two reactors would eliminate an entire plant. Second, the employment rate in their region was better than at the other plants; third, and, not least important, was Danish pressure.[98] The minister of energy, Birgitta Dahl, emphasized that the decision was "irreversible."[99] In Barsebäck plant employees and their families and friends formed a human chain around the plant as a demonstration of their wish to continue operation. The action was described as a symbolic hug.[100]

Parallel to lobbying for the continuation of nuclear power, Sydkraft began planning to replace Barsebäck with other sources of energy, namely, coal gasification and natural gas (the latter based on imports from the Soviet Union).[101] In communication with the municipality, the company emphasized employment security in the local area. Indeed, it claimed that if a coal plant were to be built instead, the number of local employees would actually increase, at least during the startup phase.

However, Sydkraft's plan for a coal gasification plant was met with fierce resistance among the local population.[102] A coal plant was considered dirty in comparison to "clean" nuclear power, and implied a number of elements that were regarded negatively, such as emissions and pollution from the plant, from coal transport trucks, and from a planned coal storage field in the area. The protests became loud and were highly organized, putting local politicians under intense pressure.[103] Kävlinge's municipal executive board suffered especially, since this body had the right to veto the project by refusing to grant Sydkraft the necessary building license.[104] In order to manage the delicate issue, a special group made up of representatives from all political parties was formed. Ultimately the municipality decided to reject the Sydkraft application for a new coal-fired power plant at Barsebäck.[105]

Simultaneously, the deep internal conflict between antinuclear groups and trade unions in the national Social Democratic Party intensified and was now played out in public. Eventually, the minister of energy, Birgitta Dahl, was forced to step aside. In 1990, the "irreversible" decision to close down two reactors was rescinded and slightly later a kind of compromise was reached, based on investments in renewable energy and, for the time being, continued operations at Barsebäck.[106] However, the pressure on the Swedish government from Denmark to shut down Barsebäck grew during the 1990s.[107]

From the area of popular culture, a strong but nonflattering image of the Swedes in relation to nuclear power was presented in a monologue from a TV series, "Riget," produced by Danish director Lars von Trier, which was aired in 1994. One of the main characters, a Swedish chief surgeon played by actor Ernst-Hugo Järegård, stands on the roof of the big hospital in Copenhagen, looking at Barsebäck, which he can discern at a distance on the Swedish coast. He sees the chimneys of Barsebäck as watchtowers and expresses his severe disgust for Denmark and for the Danes: "Thank you, Swedish watchtowers. By plutonium we bring the Danes to their knees. Here: Denmark, excrement emitted from limestone and water. And there: Sweden, cut in granite. Danish bastards. Danish bastards!"[108]

In truth, the Swedish attitude was far from conciliatory, as became clear when five of the Swedish nuclear power plants went offline for several months because of a serious technical malfunction at Barsebäck in 1992. To compensate for the standstill, Sydkraft imported electricity from coal-fired power plants in Denmark.[109] Exchanging nuclear energy for coal energy was much criticized by residents of southern Sweden, who considered it a bad solution for the environment in general, all the more so since under normal wind conditions the sulfur emissions from Danish coal plants were blown right in their direction.[110] Swedish debaters accused the Danes of using Barsebäck as an alibi to avoid their own energy problems, especially their "furious investment in coal," their too liberal environmental legislation, and also their reprehensible import of coal from apartheid South Africa.[111] An employee at the Barsebäck nuclear power plant was not alone in his view when he stated: "We are confounded with the Danes who belch out their bloody coal power smoke and there are always western winds...and it lands here with acidification problems...which destroy the environment in the long run."[112]

The friendly cooperative spirit that initially marked the relationship between Sweden and Denmark in the issue of nuclear power thus changed dramatically. Not only did young OOA activists raise their voices against Barsebäck, the Danish government also expressed severe discontent with the location of a nuclear power plant just about twenty kilometers from the Danish capital of Copenhagen. The antinuclear winds in Sweden also had a quite concrete effect on the future prospects of the Barsebäck plant, its employees, and the surrounding area. The response from the local community in Löddeköpinge/Kävlinge was nevertheless affirmative toward a continuation of nuclear power, illustrated by, among other things, the intense rejection of a coal gasification plant as an alternative source of energy and an alternative workplace.

The Spirit of Barsebäck

"We fought against odds...we will show them!"[113] "It is a great workplace; everybody is so nice and helpful."[114] There seem to be two interrelated meanings of "the spirit of Barsebäck" as it was used by the plant employees. One denoted the collective resistance toward antinuclear activists and political decisions that went against a future for nuclear power in Sweden. The other aspect was that the plant was a highly appreciated workplace where people stayed a long time. Taken together, these two meanings invoke an understanding of a workplace where employees were quite content with their work and closely united against a doubting world outside, against the "external enemy."[115]

When the two reactors of Barsebäck were online, the plant employed about 350 people.[116] A majority were men working with technical issues. About 25 percent were women, primarily performing administrative tasks like bookkeeping, personnel, and providing information to the public.[117] A large proportion of the employees worked at Barsebäck for a long time, often beginning their career in the basic operations and then moving on to other divisions of the plant.[118] Even the director as of 2010, Leif Öst, once began "on the floor" to end up at the highest position at the plant 38 years later.[119] The staff turnover was very low and the average employment time 17–20 years. A saying was that you were a newcomer if you had not yet worked ten years at Barsebäck, and it was never difficult to find new personnel if needed.[120] Many of the employees pinpoint the possibility to advance rapidly within the company as one of the reasons why they chose to stay for such a long time. Other advantages were a nonhierarchical atmosphere and the feeling of belonging to a family.[121]

Nevertheless, the pressure from outside was intense and many employees had to take a conscious position regarding their workplace. They had to justify their choice of workplace at private parties, for example, and be prepared to debate the pros and cons of nuclear power at any time. When they went to work, the road was sometimes blocked by antinuclear demonstrators calling the plant employees "child murder[ers] and such things."[122]

The employees developed a relaxed approach to the political pressure. For example, a man recalled the turbulence when on one occasion Greenpeace entered the fenced area, climbed the roof of one of the reactor buildings, and attached a huge banner on the wall: "It was like when the Eurovision Song Contest was on television, the telephones rang incessantly."[123] This man worked as a guard at Barsebäck for many years and related his expectations for weekends during periods when protests were intense: "When you came in Friday and were to work Friday, Saturday and Sunday, you were one hundred percent sure that something would happen... It has been fun in that way... it has always happened a lot."[124]

In the communications between the four reactor sites in Sweden, it was a friendly joke that the other three were happy that Barsebäck "took care of" all the negative publicity on nuclear power in the country. The public information officer at Barsebäck claimed that there was probably no other workplace where so many people had such extensive media experience.[125] The "we" and a "them" from the perspective of the plant employees were thus—at least as expressed by these quotations—created in a good-humored way, patronizing but not hostile toward antinuclear activists. Similarly, in its company chronicle Sydkraft characterized the big antinuclear marches as public entertainment, a mixture of a march and "old-fashioned school field trips to the beech forest."[126] In spite of the easygoing accounts of intense antinuclear protests, it was also stated that "there is probably no other industry...that has been so ridiculed and taunted." The response from the nuclear plant employees toward the mistrust from outside the nuclear sector was to "show that the technology works."[127]

Accordingly, the special features of nuclear power were played down by the plant employees, who emphasized the ordinary and the normal. Barsebäck was labeled an "electricity factory," and the possibility to visit the plant was regarded an important tool to show that nuclear power was almost just like any industry, only very clean.[128] Some Danish visitors were described as really scared, "frightened of Barsebäck from childhood," but after the visit they were said not to be afraid anymore.[129] By being accessible and showing visitors that there was nothing secret or hidden at the plant, the Barsebäck employees perceived they often managed to stop the fear. One employee, Jan Pålsson, stated: "When they come here and realize that...it is not anything more special than a way to boil water...they lose this fear of it."[130]

One could, of course, question the way plant employees demonstrated the safety of nuclear power. In her study of the nuclear fuel reprocessing plant La Hague in France, anthropologist Françoise Zonabend depicts an encompassing but suppressed anxiety about nuclear technology, among employees as well as among those living in the neighborhood. She shows how fear was muted and concern banished by the use of different strategies, for example, by simple denial, by drawing on tradition, or by comparison with risks in other work places and in everyday life, like driving a car.[131] Zonabend also pinpoints the habit of using vocabulary from the domestic sphere—cookery and housework—as a way to play down or to trivialize danger at the La Hague plant.[132] However, it is worth noting that Zonabend carried out her fieldwork in the late 1980s, that is, immediately after the Chernobyl explosion. It cannot be ruled out that the temporal proximity to these events may have had some effect on the responses received in the study.

The indicators I found in relation to Barsebäck, ranging from statements and actions by the municipality, local results in the referendum on nuclear power, and opinion surveys, to interviews with employees and plant neighbors, all point in the same direction: people living in the vicinity of Barsebäck feared the plant less than people living further away.[133] In chapter 4 about the Ignalina nuclear power plant in Lithuania, we will discern a similar picture. Neither at Barsebäck nor at Ignalina does the distance of fear seem to have changed much over time.

Nevertheless, there was a feeling of bitterness in the air at Barsebäck. Employees and plant management described how the plant produced cheap and clean electricity in an environmentally friendly, safe way for decades and did not release anything but some hot water.[134] In addition, they highlighted how the nuclear sector carried a huge responsibility for its business as compared to other sectors—from the cradle to the grave—and Barsebäck's management assumed that no other industry in the country would be able to meet the same demands.[135] In spite of these circumstances, very little credit was given for their dedicated efforts—only the repeated view that nuclear power was dangerous, along with the continuous threat of closing down the plant.[136]

Closing Down

"We have always lived with the threat of closing down; ultimately, you did not believe in it."[137] Over what was often a long career at the Barsebäck plant employees got used to the fact that the future of their workplace was continuously debated and questioned, and that the plant would possibly be shut down before its technical lifespan had ended. In 1997, the Social Democratic Party, the Left Party, and the Center Party eventually agreed to close down Barsebäck's two reactors and the decision became reality in 1999 and 2005, respectively. At this time, Sydkraft, like the rest of the Swedish power industry, had been increasingly internationalized. In 2001, German E.ON bought the majority of the shares in Sydkraft and a few years later, Sydkraft was named E.ON. As compensation for having to close down the Barsebäck nuclear power plant for political reasons, E.ON was granted shares in another Swedish nuclear power plant, Ringhals.

When the closure at last became real, the employees characterized the decision in terms like "madness, capital destruction, political games and intrigues, innocent victims, betrayal."[138] Many quit their jobs as soon as the first reactor closed down, since it became "too emotional and tough to be completely disregarded."[139] Representatives of the town of Kävlinge articulated an experience of being abandoned, of disillusionment in relation to the state, bemoaning that nobody outside the municipality "fought for us."[140]

E.ON put together a generous conversion program for the personnel at the plant. Initially it comprised an employment guarantee, ensuring that the plant would not experience a massive flight of competence when the closing date was announced. Later the program offered possibilities to study or to try another job at full salary for three to five years. While this helped a considerable number of employees find new ways to earn a living, other employees who became redundant found employment at other nuclear power plants in the country. After some years of adjustment, the disappointment and bitterness at the plant faded away and the situation was more or less accepted.[141] The atmosphere at Barsebäck changed into cautious optimism, and employees found new meaning in meeting the difficult challenge of decommissioning a nuclear power plant. Thus, in spite of being a "closing down industry," in many ways the enterprise succeeded in keeping the "Barsebäck spirit" alive, now with a focus on decommissioning the plant in the best possible way.

In Denmark, the OOA activists had reoriented their efforts toward the broader issues of energy politics after the 1985 parliamentary decision not to build commercial nuclear power plants in Denmark, but the organization did not formally disband until 2000 following the closure of the first reactor at Barsebäck.[142] The presence of Barsebäck in the Danish debate was not over, however. For example, in connection with the events of 9/11 in New York, a Danish newspaper featured a spread on Barsebäck as the worst terror target, combined with a montage showing an airplane on its way to hit the nuclear power plant. The short two-minute flight from Copenhagen airport to Barsebäck was emphasized and the text referred to a major who asserted that crashing into Barsebäck would be easier, cheaper, and more efficient than any other biological or chemical weapon.[143] One of two headlines stated that "We are Defenseless" and showed a picture of five people wearing full-body radiation suits.[144] The state of alert at the Swedish nuclear power plants was also raised, and the director for a Swedish nuclear expert authority said that it was practically impossible to protect oneself from an attack like the one in the United States, as modern nuclear power plants are not built to resist a suicide attack of that kind.[145] In several respects, Barsebäck was considered a "Danish" nuclear power plant, only located on the other side of the Sound; for many people the Danish nuclear era did not end until 2005, when Barsebäck's second reactor was finally closed down.[146]

Significance and Future Use

"Nobody knew where Kävlinge was located, but if I mentioned that Barsebäck was in the municipality, everybody knew."[147] The significance of the nuclear enterprise in Kävlinge municipality was huge in terms of identification and in the many ways in which the plant was present in everyday life—ranging from investments in road infrastructure, sponsorship of sports clubs, generous plant tours for schoolchildren, and distribution of special alarm equipment for the homes, to visually dominating the horizon from almost anywhere. The plant constituted one of the major employers in the municipality, but was not the only one, and within commuting distance there were a large number of other work opportunities.[148] That is, the local community was by no means as dependent on the plant as many mono-industrial towns are on their industrial enterprise. However, as illustrated by the quote at the start of this paragraph, in the logic of putting a place on the map, the name of Barsebäck held a top position in Swedish and Danish consciousness.

For the nuclear sector in general, Barsebäck was a kind of nursery where many people started their career. As an E.ON representative expressed it: "That's where it began [at Barsebäck]"; people who are now working in different positions in some of the nuclear enterprises or authorities in the country "know each other through Barsebäck, it is special."[149] The significance of what the plant produced is also an important part of the conceptualization. The plant production is compared to the amount of electricity produced by the rivers of northern Sweden, and its location in the southern part of the country is a key feature—as a unique and valuable southern input point to the power grid.[150] The negative image is also mentioned by plant employees, that Barsebäck "happened to be the symbol of antinuclear activities in the country."[151]

The silhouette of Barsebäck is certainly visible from a long distance, but this subject is worth some elaboration. During the planning and construction process in the late 1960s and early 1970s, the monumentality of the plant and its expected visual dominance in the landscape was much emphasized by participating architects, engineers, and heritage organizations as something good and desirable—but perhaps also as something that people needed to get used to before fully appreciating it. The contradictory idea that the gray color and black vertical stripes would merge with the horizon and turn into a kind of camouflage is a story that was retold by some interviewees—with a laugh: "I know that the only thing you see from the bridge [crossing the Sound between Sweden and Denmark] or from Copenhagen is...Barsebäck."[152]

The visibility from Copenhagen has been an often repeated circumstance used to illustrate the short distance between the plant and the Danish capital. What you can see is not far away, at least when it comes to nuclear power plants, and especially when a majority of those living in Denmark with the view of Barsebäck at a distance carried strong antinuclear feelings. In striking contrast, it seems as if the idea of letting the plant disappear into the landscape has been realized in the local area, although not primarily because of the gray color and stripes. According to a study from 2010, people living close to Barsebäck do not "see" the plant since it has actually "merged with the landscape" in some sense.[153] It is just there, not anything that is noted. The visual dominance of Barsebäck is perhaps more prominent in the symbolic mental gallery of many Swedes and Danes living further away from the plant, based on newspaper photos and broadcast features with national coverage, but certainly also on the Danish view over the Sound.

Dominating the landscape or not—how is the future of the Barsebäck site envisioned in the context of the ongoing decommissioning? And have there been any heritage activities connected to the plant that could affect these plans and conceptions?

In the late 1980s, two heritage professionals, Kristian Berg and Lars-Eric Jönsson, argued that the Barsebäck nuclear power plant ought to be considered from a heritage point of view.[154] Their plea was regarded mostly as a joke, however, and the issue never entered the formal agenda of the heritage authorities. Fifteen years later, in the period between the closure of the first and the second reactors, the regional museum of southern Sweden initiated a seminar about the possible heritage values of Barsebäck and also carried out a documentation project based on perspectives of building conservation and ethnology.[155] At the national level, too, nuclear power attracted heritage attention at the beginning of the twenty-first century. The smaller reactor in Ågesta that had provided district heating to the Stockholm suburb of Farsta was documented by museum professionals, and the National Museum of Science and Technology published a book on the nuclear issue in a historical perspective.[156]

Hence, it was not very provocative when Barbro Mellander, director of the regional museum, stated in 2006 that Barsebäck is "living cultural heritage in our collective memory."[157] The plant was said to fulfill the criteria used by the heritage institutions to define building structures of high cultural history value and to mirror the "grand narrative about Swedish energy politics, environmental politics and the welfare society after the Second World War," in many respects the same heritage values that were articulated in the late 1980s.[158] It seemed as if Barsebäck was well on its way to receiving heritage status, within a national or southern Swedish regional context.

In Denmark, when OOA disbanded, both the organization itself and the National Museum of Denmark encouraged antinuclear activists to write down their memories.[159] Among the answers we find, for example, former activist Helle Green, who ends her story by saying that "I have no doubts that if we had not fought in the 70s and 80s, we would have nuclear power in Denmark today. And I have a clear conscience when I think about myself taking part in that struggle."[160] The National Museum furthermore collected artifacts such as badges, banners, and publications from the OOA, but also things like a sculpture made from a piece of iron from the Swedish ship "Sigyn," which transported radioactive material from all the Swedish nuclear power plants. Sigyn actually ran aground just outside Barsebäck on its maiden voyage in 1982—the dented piece was removed and given to Greenpeace, which gave it to the OOA, which had it turned it into a sculpture. In 1999, the sculpture ended up in the National Museum in Copenhagen as a symbol of the decisive Swedish presence in the Danish nuclear power debate.[161] In 2001, the museum opened a new permanent exhibition entitled "Denmark Stories (1660–2000)," which contains one section about the Danish antinuclear movement, including Barsebäck (figure 3.4).[162]

In 2006 documentation and preservation efforts were also undertaken by the National Heritage Board in Denmark and the Electricity Museum, focusing on Risø as a research site and workplace. The decommissioning of the first reactor had been completed in 2005 and the reactor building left empty.[163] The documentation project in 2006 drew tremendous interest from the decommissioning company and many previous employees at Risø. The investigator asserted that because of the negative attitude toward nuclear power in Denmark, the history of the Risø site had not received the attention it deserved, neither in terms of economic and research issues nor as far as national and cultural identity were concerned.[164] Risø apparently did not fit well into the narrative of nuclear power in Denmark.

Even though it no longer produces any electricity, many visitors find their way to the Barsebäck plant, in what is sometimes labeled

Figure 3.4 At the Danish National Museum in Copenhagen, visitors can learn about the history of the antinuclear movement. Among the key symbols are the Swedish nuclear power plant Barsebäck and the "sun badge," designed by a young Danish activist in 1975 and translated into 42 different languages. Photo: Anna Storm, 2010.

nuclear power tourism.[165] The information officer states that many people nowadays find closed down factories exciting in general, and that Barsebäck is even more exciting, almost a bit "mythical," since it is a "symbol for nuclear power."[166]

Nevertheless, for the time being Barsebäck's owner E.ON is not planning to preserve anything of the physical structure of the plant, nor anything of the archival material beyond what is required by law.[167] Some employees at Barsebäck think the plant, or parts of it, could be an important industrial memory to preserve, while others think it would be a waste of societal resources: "There is nothing beautiful or cultural about a nuclear power plant. It has the right to exist only as long as it does its job."[168] In line with the latter statement, the company's goal is that the plant should disappear as completely as possible, and that the site could be used freely for other purposes. Whether or not a radioactive site can be regarded as "clean" naturally depends on whether the future use includes a residential area where vegetable gardens will be planted, or whether it remains a fenced-in site for industrial production.[169] Representatives of Kävlinge municipality wish to see a new high-status residential area take shape along the coastline with the view over the Sound—"Barsebäcks sjöstad" (Seaside Barsebäck)—assuming that the ground can be satisfactorily decontaminated.[170]

A heritage perspective on the plant has mostly been a nonissue from the municipal point of view as well as for E.ON, and the heritage institutions have also silently withdrawn.[171] As of 2013, there are no signs of heritage interest and no contacts between the plant management and the regional museum or any other heritage institution.

Concluding Remarks

Close by the Barsebäck plant there is a mound of stones built by antinuclear activists in the 1980s. If the plant is completely taken away, the mound will be the only remaining material witness of the nuclear power plant at the site.[172] The plant is still there as a reminder, a monument that evokes strong but conflicting feelings. Will it be possible for the scars in the landscape to heal if the physical reminder disappears? By itself, the mound is an insufficient foundation to bear the memories connected with Barsebäck.

Barsebäck nuclear power plant became an esteemed workplace and societal actor in the local context of Löddeköpinge and Kävlinge in southern Sweden. It was symbolically defining since everybody in the country knew about the plant, and at the same time, it turned "invisible" and become normalized by those living close by. From a national point of view, Barsebäck was the foremost emblem for long and distressing political controversies about the future of nuclear power in Sweden, not least since the plant was a cause for tense relations with neighboring Denmark. In Denmark, Barsebäck was almost considered a domestic plant, only located on the other side of the Sound, and as such, first something to encourage and later to fight.

Just like in Malmberget, the physical landscape reminiscent of these critical decades is planned to disappear as completely as possible, and this ambition is not contested, but rather the opposite. Reinstating a "natural" landscape is the current benchmark for responsibility in nuclear matters. The distance of fear, the scales that decisively affect the significance of a particular nuclear power plant, will therefore have to attend to a new, less distinct geography in the future.

Chapter 4
Lost Utopia

The dream of the Soviet paradise had many components. Men and women were to be equal, and different duties were to represent the same value. Work and leisure were to be smoothly combined and the living environment to comprise both the city and the countryside. During the Soviet era, new settlements were established on previously uninhabited land, moving people over great distances to create new ideal worlds based on industrial enterprises. Did the dreamt paradise ever come close to realization anywhere within the vast empire? And what happened to the dream when the empire collapsed in 1990?

The post-industrial scar of this chapter tells of dreams and betrayals, tensions created by ethnicity and social status, and risks that were concealed by privileged living conditions. The place is the Ignalina nuclear power plant, built in the Soviet republic of Lithuania in the 1970s and 1980s, and the adjacent workers' town of Sniečkus/Visaginas (map 4.1). The scar is defined by the changing state borders of 1990, which placed the nuclear engineers and the town's population in different kinds of geographical peripheries and also shaped the significance of the nuclear power plant from the various viewpoints of Moscow, Vilnius, and Brussels.

The Ignalina plant was once the largest nuclear power plant in the world and operated for about twenty-five years. During these years, the plant made a remarkable symbolic journey from being an expression of advanced Soviet technological progress, to a despised sign of Moscow's imperialistic ambitions threatening human health, the environment, and ultimately the Lithuanian nation, to prove to be a most valuable energy resource for independent Lithuania in its relations with Russia and the West and, finally, to end its operational life as a concession to the negotiations concerning Lithuanian EU membership. Left at the plant were radioactive waste and a town full

Map 4.1 The Ignalina nuclear power plant was built in the Soviet republic of Lithuania, close to its borders with Latvia and Belarus. A completely newly built workers' town, Sniečkus, was established six kilometers from the plant. When the Soviet Union collapsed, the town was renamed Visaginas and is now located on the periphery of independent Lithuania. Map: Stig Söderlind, 2013.

of people seeking new ways to earn a living. The Ignalina plant did not experience the horrors of a nuclear accident as in Chernobyl; very few nuclear tourists and antinuclear activists have found their way to the former "queen" of the Soviet nuclear world.[1]

The Plant and the Town

The Ignalina nuclear power plant was scheduled to consist of four so-called RBMK reactors, each with a nominal electrical output of fifteen hundred megawatts.[2] Even when just two reactors had been completed, in 1983 and 1987, Ignalina was already the most powerful nuclear power plant in the world. During construction of the plant, a workers' town named Sniečkus was built nearby, designed to accommodate thirty thousand people.

In the Soviet Union nuclear matters were the purview of the central government.[3] The secret and powerful Committee on Atomic Energy, reporting directly to the Ministry of Medium-Scale Machines (*Minsredmash*), decided and directed what was to be built when and where, and the purpose of the Ignalina plant was to increase the production of electricity in the northwestern part of the Soviet empire. Two locations in the republic of Belarus and one in the republic of Lithuania were considered and the choice fell upon the Lithuanian site, because of its better soil, the large Lake Drūkšiai that could be used for reactor cooling purposes, and the existence of some infrastructure, primarily a railway.[4] In spite of the railway, the area was sparsely populated, with smaller farmsteads in a landscape of forests and lakes. The scenic region was a favorite vacation destination.

Construction began in 1978 and involved a total of twenty-two thousand people. Skilled nuclear engineers were recruited from different parts of the Soviet Union along with builders and less-qualified workers, some of them sent by the Soviet Army.[5] The living and working conditions were hard during these first years, but a pioneering spirit and the feeling of being chosen for this important task on the Soviet "frontier" helped considerably in withstanding the hardships.[6] One woman remembers that in "the beginning there were only forests and marshes and there came the best people of the Soviet Union, devoted to their country, to build the plant and the town."[7] Another story illustrates how it was considered a privilege to move to this part of the empire, recalling the first arrival in Sniečkus: "We were driving, and the fields were green, such beauty! I was thinking—oh God, what a place!"[8]

The town of Sniečkus was built according to a general planning concept of mono-industrial towns—urban structures built exclusively to provide a specific enterprise with a workforce.[9] Often these towns were located in previously uninhabited areas and became representations of a Soviet utopia—to join work and leisure, city and countryside, with an emphasis on equality between men and women and between different occupational groups.[10] Since the state owned the land, space was never a problem, and Sniečkus became characterized by fresh air and greenery, yet following the so-called eight-minute principle, meaning that one could reach the center from any part of

Figure 4.1 The mono-industrial town of Sniečkus has in retrospect been labeled an "elite version" of Soviet planned space. In the center of the picture is a stone with a plaque commemorating the foundation of the "atomic city." Courtesy of the Energy and Technology Museum, Vilnius.

the town within this period of time. In retrospect, Sniečkus has even been called a "Soviet paradise" (figure 4.1).[11]

To these general characteristics of mono-industrial towns was sometimes added the element of living in a closed or semiclosed town, depending on the level of secrecy surrounding the industrial activity. Sniečkus was a semiclosed town, which meant that entry was free but becoming a permanent resident was controlled.[12] The inconveniences of living in a restricted town were compensated for by comparatively high salaries, a better standard of housing, and access to what were regarded as luxury goods, which is why these towns are sometimes referred to as "chocolate cities."[13] The enterprise—in this case the nuclear power plant—and the town were intimately intertwined. Among those employed at the plant were also people providing social services and the infrastructure of the town, such as housing, kindergartens, schools, hospitals. Consequently, when the total number of employees at Ignalina was about sixty-five hundred, only thirty-five hundred of them took the free commuter bus the six kilometers from Sniečkus to work at the actual plant.[14]

Despite the ideal of equality, the status of the skilled nuclear engineers and physicists was high and noticeable in terms of even better

Figure 4.2 To work at the Ignalina nuclear power plant was a well-paid, high-status occupation for highly educated physicists and engineers. The picture shows the control room of one of the reactors. Photo: A. Karasev. Courtesy of the Energy and Technology Museum, Vilnius.

salaries and social benefits (figure 4.2). They were the most prominent of the local Soviet "heroes" and well aware of their importance. Besides the social differences shaped by occupation, anthropologist Kristina Šliavaitė discerns how tension between groups of people in Sniečkus emerged according to ethnicity and rootedness.[15] Since Sniečkus and the Ignalina plant were built on previously "untouched land" (or rather by dismantling 165 farmsteads) a vast majority of the inhabitants had moved there, most of them from distant places.[16] The common denominator was the Russian language, and it has been argued that the establishment of mono-industrial towns in peripheral regions of the Soviet Union was a conscious strategy to promote cultural integration.[17] A bit paradoxically, this might be an explanation for the Ignalina plant management's practice of discouraging ethnic Lithuanians from participating in the construction work and later seeking employment at the plant.[18] In the late 1970s, less than 6 percent of Sniečkus's inhabitants were ethnic Lithuanians.[19] Both the Russian-speakers and the Lithuanians were prejudiced against each other, among other things due to the clash between urban and rural identities and between different levels of education.

The name Ignalina was picked from a village located about forty kilometers from the site where the nuclear power plant was built. The name of Sniečkus was a tribute to a long-standing communist leader in Lithuania, Antanas Sniečkus.[20] The naming of the town was one of the issues that the Soviet central planning left to local authorities, along with some parts of the architectural work of the town. The Leningrad-based institute VNIPIET (All-Union Scientific Research and Design Institute for Energy Technologies), which had previous experience in planning "atomic cities," drew up the master plan of the town and—after indignant voices raised from local firms about the importance of "local color"—left the elaboration of public spaces to Lithuanian architects. Art historians Andis Cinis, Marija Drėmaitė, and Mart Kalm show how this compromise resulted in a town that combined prefabricated high-rise concrete buildings with specially designed lower red brick buildings housing kindergartens and schools. In this way, Sniečkus became what they term an "elite version" of Soviet planned space.[21]

The Ignalina plant was built according to the standardized pattern of civilian nuclear enterprises emerging in great numbers at the time. Behind an entrance area with security check were offices for management and dressing rooms staffed by special personnel. Through an elevated corridor one could reach the reactor buildings, the control rooms, and the long machine hall for the turbines, which was connected to both reactors. Outside the restricted area was the canteen. Ignalina's characteristic chimneys, three pipes leaning toward each other, are jocularly said to have been designed by a woman.[22]

Soviet Nuclear Activities

Soviet nuclear activities began primarily as a military concern. In the 1940s, after the US nuclear bombs at the end of World War II, research aimed at developing Soviet nuclear weapons was intense and constituted a crucial element of the race between the two superpowers.[23] In 1954, the first nuclear reactor in the world producing electricity for a grid began operating in Obninsk, southeast of Moscow. The Obninsk reactor (called *Atom Mirny*, or "peaceful atom") was of the RBMK design developed in the Soviet Union and had a nominal electrical output of five megawatts.[24] The RBMK design derived from plutonium-producing reactors and could theoretically be used to produce plutonium for weapons, although this seems not to have happened in practice.[25]

RBMK is a Russian acronym standing for graphite-moderated channel-type reactor. It is cooled by light water and easy to scale up or down in size.[26] Perhaps because of the potential "secondary" product of plutonium, these reactors were built only within the borders of the Soviet Union. However, another domestically developed reactor type—which did not produce plutonium—the so-called VVER (acronym for water-water power reactor), was later exported in large numbers to the satellite states of the Eastern Bloc.[27]

The nuclear program of the Soviet Union was from the beginning an empire of its own. A network of educational institutes, administrative organizations, and enterprises were supervised by Minsredmash. No independent regulatory body existed, and the budget was huge.[28] Early on energy was a key issue for the imperialistic objectives of the Soviet Union, illustrated by Lenin's well-known words: "Communism equals Soviet power plus electrification of the entire country."[29] The establishment of Soviet administration in, for example, Lithuania after its incorporation in 1940, was consequently followed by an extensive electrification scheme.[30]

Lithuania, along with the other Baltic states of Latvia and Estonia, had been electrified late. When the Soviet program brought integration in a very physical sense through a high-voltage grid, it was the first time any electrical connections had crossed the borders between the three states.[31] The northwestern part of the Soviet Union was generally considered to be in need of more electricity production, so both small and large-scale power plants were established. In Lithuania there were a combined heat and power plant in Kaunas working from the late 1950s and a thermal power station in Elektrenai that started operation in the mid-1960s.[32] However, it was the Ignalina nuclear power plant that was to make the real difference, with preparatory work starting in the early 1970s. Because of its exceptionally high output—fifteen hundred megawatts per reactor—it was to become the "queen" of the whole Soviet nuclear empire.[33]

At this time, civilian nuclear power was really taking off. With the tenth five-year plan, comprising the years 1975–1980, increased use of nuclear power for electricity production was an outspoken objective.[34] A less overt goal was to replace religious beliefs with belief in the progress of science and technology.[35] For deeply Catholic Lithuania, this was perhaps a crucial issue in terms of culturally integrating its population into the Soviet Union. However, in this case the success could be disputed, since there were still about six hundred active churches in the republic in the 1980s.[36]

When the two Ignalina reactors began operation, one of them provided the Soviet republic of Lithuania with 70–80 percent of its electricity, while the other supplied the republics of Latvia, Estonia, and nearby districts of Russia.[37] In Estonia, the presence of an oil shale industry led to the development of uranium production for Soviet nuclear needs. The geographical emphasis on the civilian nuclear power program of the Soviet Union was nevertheless located in the republic of Ukraine. In the mid-1980s, there were thirteen reactors up and running within its borders, among them four at the Chernobyl plant, close to the workers' town of Pripyat.[38]

Chernobyl

The explosion in reactor number four at Chernobyl in April 1986 was a breaking point in many respects. The infallibility of Soviet technology was all at once in doubt, although the authorities tried to conceal information about the true scale and danger of the radioactive fallout.[39] Accordingly, evacuation and protection measures were inadequate, for the emergency workers sent to cover the reactor with concrete in order to stop the radioactive outflow as well as for the people living nearby. Surely, Soviet workers had been cleaning up radioactivity after accidents before; the difference this time was that it occurred while cameras were running.[40]

The four Chernobyl reactors were all RBMKs, like those at Ignalina, although with a lower output (one thousand megawatts). Before the Chernobyl accident, the technical characteristics of the RBMKs had generally been considered favorable: the comparatively low temperature in the core was regarded as positive from perspectives of safety; the construction process, which demanded components that for the most part could be gathered from other industrial sectors, made it easy to build at peripheral sites; and the possibility to refuel without shutting down the reactor implied high efficiency in production.[41] A Soviet nuclear scientist highlighted some of the crucial weaknesses of the design early on in 1965, but his objections were silenced by threats of dismissal.[42]

After Chernobyl, the hazardous technical features became painfully clear to a larger group of people: among other things the lack of containment of the reactor, the weak self-shutdown mechanism, and the vulnerability to loss of coolant.[43] Across the Soviet Union, the construction or planning of over forty RBMK reactors was halted, among them the second reactor at Ignalina, which had been very close to being commissioned.[44] As an immediate reaction, a number of safety measures were undertaken at the Ignalina plant, including improved fire protection and a lowering of the output level.[45] When the reactor finally began operating after a one-year delay, many of the usual festivities usually surrounding such an inauguration were absent.[46] The utopia of technological progress in prosperous cities at the edge of the empire was under fire.

However, in spite of the new insights into construction failings, the Soviet nuclear organization remained virtually unshaken by the Chernobyl accident.[47] Its representatives were also rather successful in convincing the worried Western world that they would take care of the problems. The three remaining reactors at Chernobyl continued to work, but most of the town of Pripyat was abandoned and a new town, Slavutych, built about fifty kilometers away.

The Chernobyl accident happened about four years before the collapse of the Soviet Union. Yet, these two major events are interlinked, both in a kind of causality and in people's experience of the all-encompassing changes that took place in the late 1980s.[48] The perestroika period encouraged new thinking—within the existing framework of the system—in order to enhance economic development. As one part of this new thinking, it became possible to form organizations that were relatively independent of the state and the Communist Party. Such organizations could certainly not deal with any subject but, among others, environmental clubs were permitted. And this is how the independence movement in Lithuania began.

The Žemyna Club and Sąjūdis

In 1985, the year before the Chernobyl accident, the central Soviet authorities planned to drill for oil along the Lithuanian coastline. To the local population, the relatively short coastline and the Curonian Spit, a narrow peninsula that stretches along the coast with impressive sand dunes and small settlements, were highly valued vacation areas. The protests became so loud that in 1986 the oil-drilling plans were abandoned, and the first seed of a Lithuanian environmental movement was sown.[49] During 1987, a small group of intellectuals continued to draw attention to other environmental risks in Lithuania, and in 1988 the so-called Žemyna club was officially founded. With the label of an environmental organization, its major object of concern became the Ignalina nuclear power plant.[50]

Among the founders were various professionals, one of them a young nuclear physicist trained in Moscow, Zigmas Vaišvila. By means of polemical articles in the newspapers, he succeeded in generating awareness of the risks of the Ignalina plant.[51] Since he knew the nuclear power processes from the inside, his arguments were difficult to ignore.[52] Yet, a full decade earlier, during the construction of the Ignalina plant, representatives of the Lithuanian Academy of Sciences had already expressed misgivings about the project. They asserted that the scale of four fifteen-hundred megawatt reactors was too big, especially in relation to the cooling capacity of Lake Drūkšiai, and criticized the lack of detailed geological studies of the site.[53] Their plea was to limit the plant to two reactors instead of four. In the beginning of the 1980s, they managed to convince the Soviet central authorities to mothball their plans for a fourth reactor (which did not become known publicly until 1987). The third was to be built, however, and construction began in 1985 (figure 4.3).[54]

While the scientists of the academy had conveyed their concerns within the system, mainly secluded from public knowledge, the articles written by representatives of the Žemyna club in the spring of 1988 reached a mass audience.[55] As the spring unfolded, the Žemyna club initiated public meetings and marches against the Ignalina nuclear power plant, and some of the scientists from the academy

Figure 4.3 In spite of protests from Lithuanian scientists, the construction of a third reactor at the Ignalina nuclear power plant began in 1985. Locals were mainly concerned about the cooling capacity of Lake Drūkšiai, and persistent protests eventually thwarted the plans for a fourth reactor. Photo: Vaclovas Kisielius. Courtesy of the Energy and Technology Museum, Vilnius.

spoke at these events.[56] Žemyna also collected signatures—forty-four thousand names expressing their worries about Ignalina—and sent them to Moscow. The authorities in Moscow took the activities going on in Lithuania seriously, but many of the local communist leaders seemed ambivalent.[57] In June 1988, the Lithuanian Movement for Perestroika, more known simply as Sąjūdis ("the Movement"), was formed, and many of the members also belonged to the Žemyna club.[58]

In September, Sąjūdis arranged a weekend-long protest meeting at the Ignalina plant, and 15,000–20,000 people spent two or three days living in tents outside the plant, and on one occasion they formed a human chain around it.[59] A priest from a nearby church led prayers and gave a short speech encouraging the activists. In spite of the size of the event and the physical closeness between the activists and the management and employees at the plant, there was no serious confrontation. One source even relates that some of the protesters took guided tours of the plant.[60]

Unsurprisingly, the residents of Sniečkus did not sympathize with the demonstrators at all.[61] On the contrary, the difference between the views and wishes of the people of Sniečkus and the visiting supporters of Sąjūdis was striking. The population of Sniečkus wanted everything to continue as it was, and the Sąjūdis supporters wanted radical change. During the summer both the Ignalina nuclear power plant and the town of Sniečkus had been depicted as threats by Sąjūdis representatives.[62] The nuclear power production was said to threaten human health and the environment, while Sniečkus was described as embodying "national contamination" and "demographic pollution."[63] The overwhelmingly Russian-speaking majority of Sniečkus was thus seen as an alien element from the perspective of an otherwise ethnically homogenous Lithuania and, thus, combined with the existence of the huge nuclear plant, as a detested symbol of Moscow's dominance.[64] Hence, opposition toward the Ignalina plant shifted in emphasis and framing. The previous focus on nuclear-related hazards was overthrown by a nationalistic concern for the survival of Lithuania. Political scientist Jane I. Dawson even terms the Lithuanian environmental movement a "substitute" for the true struggle for an independent nation, compelled by the political restrictions at the time.[65]

After the 1988 September meeting at the Ignalina nuclear power plant, Soviet authorities eventually agreed to invite the International Atomic Energy Agency (IAEA) to review the plant's safety. Sąjūdis thereafter seemed to lose interest in Ignalina and concentrated on more overt issues of Lithuanian independence.[66] A year and a half later, in March 1990, Lithuania declared its sovereignty, the first republic of the Soviet Union to claim independence. The central government headed by Mikhail Gorbachev responded, among other things, by a blockade of Lithuania. The Ignalina plant was never shut down, however, which some contemporary commentators point out as the decisive factor as to why Lithuania was able to withstand the seventy-five-day-long blockade.[67] Electricity was available, and the highly agrarian country did not have any major problems with food supply.

The series of events that took place in Lithuania in the late 1980s had its counterparts in other Soviet republics. The connection between environmental protests directed toward nuclear power plants and anti-Soviet mobilization was also evident in, for example, Ukraine and Armenia. And all of these environmental movements faded away just as quickly upon the collapse of the Soviet Union. In Lithuania the movement lasted a little more than a year, while in Armenia it lived for only a couple of weeks.[68] In addition, a firmly established experience of national identity (in Lithuania combined with an ethnically homogenous population and living memories of independence) seems to have been a common denominator for the Soviet republics that declared their sovereignty early on, for example, Latvia and Estonia.[69] In republics where this clear group identity was missing, independence came a little later.

Ignalina and Sniečkus within New Borders

It was "the borders that came over the people and not the people who came over borders."[70] This is how one of Kristina Šliavaitė's informants described the situation for the people of Sniečkus after 1990. The new borders of an independent Lithuania brought fundamental changes to the town and the plant with regard to their economy, social status, and future prospects.

In the Soviet system, the nuclear power plant had received its funding from the central organization Minsredmash, and electricity had been delivered to the users as one service among others. When the Soviet Union collapsed, electricity continued to be a sector under state control, but now the governments of the new independent republics were in charge. In other words, the production of electricity was not adjusted to the market economy conditions faced by autonomous companies.[71] However, payment for the product soon became a problem. The new customers were electricity users who had never had to pay bills for this utility, and faced with the economic constraints of the first years of the 1990s, they had little intention to start doing so now.

The result was that the Ignalina nuclear power plant continued to produce electricity but did not receive any payment for its deliveries. Before long the plant management was not able to pay salaries to employees, and issues like plant maintenance were certainly put on the back burner.[72] These circumstances obviously affected the safety level, not the least since even "highly qualified workers and skilled technicians become lax when they are not paid," as historian Paul Josephson put it.[73] In addition, the plant's previous engagement and responsibilities for social services in the town of Sniečkus were cut by the economic shortages.

For town residents the economic pressure was coupled with significant social change. Suddenly, all high-level positions in society demanded fluency in the Lithuanian language. In spite of language courses offered in the 1990s, the Russian-speaking elite of Soviet-era Sniečkus saw its career opportunities diminish considerably.[74] In Lithuania, citizenship was granted to all permanent residents, but the right of ethnic minorities was not well articulated. During the first decade of independence more than one hundred thousand Russian-speakers left the country, and Sniečkus was no exception.[75] Besides the new language politics, the general status of different groups of people shifted: those who had been the Soviet "heroes" and the "carriers of civilization" to these previously forgotten forests were suddenly negatively associated with the former imperial power.[76] Unsurprisingly, grief and nostalgia for the Soviet period were prominent in Sniečkus—the town renamed itself Visaginas in 1992.[77]

At the nuclear power plants of the former Soviet Union and in the former satellite states of the Eastern Bloc, the dependence on Russian specialists and the central nuclear organization was crucial. In the 1990s, many of these enterprises were left in a vacuum and had to start building their national nuclear expertise and institutions from scratch, as the Russian operators sometimes literally said goodbye and handed over the key.[78] In Lithuania and at the Ignalina plant, the situation never became critical since emigration was not that extensive, and, according to one former manager, the connections with Russia were not seriously interrupted.[79] One thing that vouched for prolonged contacts with Russia was the fact that the fuel to the Soviet designed reactors was centrally manufactured and, thus, in order to continue operation of the RBMKs and the VVERs, the newly independent states and the former satellite states had to stay on speaking terms with Minsredmash, or the Ministry for Atomic Energy (MinAtom), as it was renamed in 1992.[80]

Independent Lithuania established a national nuclear law, a Lithuanian energy institute, and a nuclear regulatory body called VATESI to continue operating Ignalina.[81] During the 1990s, however, the export of electricity from the Ignalina plant to destinations outside Lithuania dropped by 90 percent. Instead, the plant carried on providing the country with 70–80 percent of its electricity, coupled with a vast unused overcapacity. From the perspective of the new government of Lithuania, the potential to export electricity to Western countries became a tempting alternative.[82] A large proportion of the members of the new Lithuanian government were Sąjūdis representatives, and since many of the key actors of Sąjūdis had their background in the Žemyna club, it turned out that former Žemyna activists were now in positions of political power. The environmental movement had disappeared with the rise of the independence movement, and now the individuals previously protesting against the Ignalina plant were actually resuming expansion plans for the same very plant.[83] This was also the case in other independent republics, where moratoriums on new reactor construction from the days of the Chernobyl accident were now lifted without much protest. Nuclear power plants had become a most valuable economic and political resource for the former Soviet states.[84]

The idea to export the abundance of electricity produced at Ignalina to other countries was not easily realized. From the perspective of countries like Sweden, an increased flow of electricity over the Baltic Sea was definitely desirable—but in the other direction. The vision from the West was generally based on two objectives: first, to force a closedown of what were perceived as dangerous nuclear plants in the East and, second, to export its own electricity into the resulting open market.[85] Obviously, the respective electricity export plans were difficult to reconcile. To these future scenarios of economic and infrastructural integration between the former Soviet states and the West came a large-scale international involvement in upgrading safety at RBMK and VVER reactors.

Improved Nuclear Safety and a Sacrifice to the West

The accident at the Three Mile Island nuclear power plant in the United States in 1979 prompted some new international cooperation in the nuclear safety business. Mostly, however, it resulted in national reviews of domestic plants to assess how these might be improved.

With the accident at the Chernobyl plant international engagement intensified, but it was the collapse of the Soviet Union that made national nuclear associations begin to consider treating the field as a matter of serious international concern.[86] Visits to nuclear power plants in the former Soviet Union shocked Western observers. One report told about "loose cables and wires...the reactor hall was scattered with waste...contaminated equipment was discarded in a corner."[87] In the first years of the 1990s, international conferences and initiatives were held in rapid succession. The European Bank for Reconstruction and Development (EBRD) was formed as a financial means to support the former Eastern Bloc in general and became a key funder in nuclear safety matters. The European Union acted as well, through the European Atomic Energy Community (Euratom).[88]

Besides the organizations with a wider international outreach, manifold bilateral and multilateral programs were launched. Within this massive amount of foreign investment the Ignalina plant was to become "unique in the scope and comprehensiveness of the safety analyses that [were] conducted" there.[89] First on the spot was Sweden. Ignalina was geographically closer to Sweden than Chernobyl, and the Chernobyl accident had affected several areas of Sweden with radioactive fallout. Therefore, the Swedish government was eager to invest money in safety measures at Ignalina, initiating what was called the Barselina project in 1992.[90]

The name "Barselina" was a combination of the Swedish nuclear power plant Barsebäck and Ignalina. It was framed as a cooperation project in which the management at the Swedish plant was to serve as partner to the Ignalina enterprise. In the first phase of Barselina, Russia also took part as a third collaborator. The experiences gained from the Barselina project were later implemented at some other nuclear power plants with RBMK reactors in Russia.[91]

The basis of the Barselina cooperation was a so-called probabilistic safety analysis (PSA), with a first stage carried out between 1992 and 1996 and a second stage running up to 2001.[92] The first stage focused on technical failures that could cause damage to the reactor core. The second stage followed up on the results from the first, estimating the potential release of radioactive material that could result from such failures.[93] The special characteristics of the PSA compared to earlier safety improvement methods was the perspective of probability rather than determinism, that is, an attempt to cover the outcomes of any technical malfunction instead of preventing accidents caused by faulty materials and components. The PSA method was comparatively new and made possible by the dramatic increase of computing power in the 1980s. The approach lacked the dimension of human failure, however, and its connection to social and economic pressures, but to some extent these aspects were covered by other approaches.[94]

For many of the individuals involved, the Barselina project became an important part of their working life, and it also brought about longstanding friendships.[95] Swedes tell about the feeling of contributing in a crucial way to improving on the poor conditions at the Ignalina plant, and at the same time express deep respect for the knowledge of the nuclear engineers they met. The managing director at Barsebäck, Leif Öst, describing a visit to the Ignalina plant, relates that "certainly it felt unpleasant to stand on top of the reactor with the steam being released under our shoes, it should not do that," but he also emphasized that the personnel was competent indeed: "They were doctors of engineering and so on, almost all of them who managed that plant, and even if we have a good level of education here, too [at Barsebäck], we do not have *that* kind of standard."[96]

Gunnar Johansson, a Swedish nuclear safety expert, recalls how the buildings themselves were a "rush job by military servicemen," but when he looked at the "pipes, pumps and valves, it was good stuff."[97] Management representatives at the Ignalina plant tell about good cooperation with the Swedes and express how grateful they were for the support; at the same time, employees at the plant explained that people coming from the West might find Ignalina dirty and dangerous, but that all depended on what you were used to.[98] It has been argued that the technical differences between Western and Eastern nuclear power plants should not be overemphasized, resulting in a simplified picture of Western reactors as safe and Eastern reactors as unsafe.[99] However, Western ideas about safety culture, that is, the human factor in safety issues, was in general not easily received by the Ignalina plant management.[100]

With the safety improvements of Barselina, along with numerous big and small projects involving other countries and bodies like the EBRD, an international safety analysis report of 1996 concluded that the Ignalina nuclear power plant did *not* have to close down immediately.[101] One of the key individuals representing Sweden, Jan Nistad, remembers how representatives for the European Union became angry with him when he refused to report that the Ignalina reactors had to be instantly shut down.[102] However, the safety analysis report was found to contain several weaknesses and so the safety improvement work continued.[103] In a general depiction of the foreign involvement in upgrading the safety of Eastern Bloc nuclear power plants, energy scholar and policy debater Nicole Foss asserts that a "prevailing assumption amongst donors that a Western approach is objectively correct and therefore universally applicable has been problematic."[104] She also assumes that Western nuclear contractors who were involved in assessing nuclear risk were interested in finding new markets for their products and therefore recommended upgrades or replacements despite enormous costs.[105]

Nevertheless, in order to receive economical support from the EBRD and others, Lithuania had to agree to certain conditions, among them that the nuclear power plant would not continue operation beyond the time the reactor channels had to be changed, which usually happens after 20 years and thus implied the beginning of the twenty-first century.[106] Furthermore, in spite of the documented safety improvements, the shutdown of Eastern nuclear power plants remained at the center of the international political agenda. When Lithuania began negotiating for EU membership, the future of the Ignalina plant became the hardest nut to crack.[107] Pleas and complaints and nightmare scenarios were articulated by those who wished to prolong its operating life, and representatives of the Lithuanian government, of the nuclear engineering education in Kaunas, of the Lithuanian Energy Institute, and certainly the Ignalina plant management, drew attention to the consequences of a shutdown.[108] They insisted that the plant was indeed safe to operate for at least another fifteen–twenty years, pointing out how Lithuania would become dependent on Russian natural gas in the case of a shutdown, how the country would lose its competence in nuclear engineering, and how Lithuania would ultimately find it difficult to stay economically and politically independent. Frustration ran deep. "Who can persuade our citizens that such a measure [to shut down Ignalina] will be beneficial and that they will enjoy a better quality of life thereafter?" asked Jonas Gylys, professor at Kaunas University of Technology, and Leonas Ašmantas, former energy minister in the Lithuanian government.[109]

In the town of Visaginas, too, many people were worried about this next step in the decline of their "nuclear way of living."[110] As Kristina Šliavaitė points out, the prospect of decommissioning Ignalina was a "final devaluation of the Soviet modernization project" from the perspective of the local residents.[111]

Eventually, all the appeals were in vain. When Lithuania entered the European Union in 2004, one of Ignalina's reactors was closed down right away and the other was slated to be closed down before 2010. From the local perspective this decision was felt to be yet another betrayal.[112] First came the collapse of the Soviet Union, and now this final kiss of death for the community. The closure was perceived as irrational and based not on technical considerations but only on politics. It was seen as an unfair decision and a waste of skills.[113]

Nuclear Fear and a Long-Lived Heritage of Radioactive Waste

Was anyone living close to the plant afraid of Ignalina? After all, nuclear hazards and radioactivity occupy a key role in public fear in general.[114] Nevertheless, testimonies of fear are difficult to find in nuclear cities like Sniečkus/Visaginas and among Ignalina plant employees. State authorities were generally regarded to be reliable and this trust in official statements brought a feeling of security to the local community. In the late 1990s, about 50 percent of the population in Lithuania and 70 percent in Latvia reported psychological discomfort due to the Ignalina plant, while in Visaginas, the figure was only 6 percent.[115] At the same time, although a majority in the greater Ignalina region considered the plant dangerous, only a minority of the inhabitants of Visaginas did so.[116] These figures thus strengthen the idea of a geographical distance of fear, suggested in chapter 3 about Barsebäck. In contrast to Barsebäck, however, the population of Visaginas was quite dependent on their nuclear power plant. In the case of Visaginas, the fear probably relied to some extent on other employment options. If you do not have much of a choice, you do not worry too much either.[117]

Radioactivity was certainly known to be dangerous, but was believed to be manageable nevertheless.[118] One nuclear specialist working at the Ignalina plant claimed that "radioactivity is like dirt, not dangerous if you are intelligent and take all the necessary precautions and behave in accordance to regulations."[119] Among the plant employees there was also a line of argument that asserted that their professionalism was of the highest level, not like those who worked at Chernobyl.[120] In this way they created a distance to the explosion and connected it to human failures, not primarily to weaknesses in the RBMK reactor type, which the Ignalina and the Chernobyl plants had in common.[121] The strong trust in nuclear energy of the early days, the 1950s and 1960s, when working reactors were displayed close to the public at industrial exhibitions, was gone, but not replaced by a general doubt in the technology.[122] In addition, the extreme exposure to radiation that had caused almost immediate devastating effects on human health and the environment at early nuclear sites in the Soviet Union was certainly not present at Ignalina (figure 4.4).[123]

Figure 4.4 In the town center of Visaginas is a pedestal with the town symbol, a crane, on top. Below is an electronic sign that alternately shows the current temperature, time, and level of radiation. Photo: Anna Storm, 2010.

There are other claims to be found: for example, an assertion by a Visaginas resident that the relatively high salaries at the Ignalina plant were legitimate because these workers were exposed to risk and performed unhealthy work, or local warnings to avoid mushrooms since they retained radioactivity longer than other plants, but these voices represented a minority.[124]

From the planning perspective, locating workers' cities just a few kilometers from nuclear plants seems to be a spatial sign of unwavering confidence in the technology, or is this just an illusion?[125] Based on studies of plutonium cities of the 1950s in the Soviet Union and the United States, historian Kate Brown argues that these workers' settlements were built as prosperous "Edens" in the midst of otherwise poor areas. The loyalty of the inhabitants was bought by an overwhelming life of everyday prosperity, manifested in the housing, food supply, education, health care, cultural activities, and so on.[126] Brown also asserts that the secrecy surrounding plutonium sites in both superpowers primarily worked to prevent knowledge about danger and risk from reaching the public, even if they were aware of the consequences. Intelligence services during the Cold War meant that nuclear activities were no secret at the governmental levels of the superpowers, rather, it was the workers and nearby residents in both the Soviet Union and the United States who had to be kept ignorant, loyal, and in a good mood.[127]

Life is never without risk; it is the level of acceptable risks that is negotiated and changing. Many considered the risk of working at and living next to a nuclear power plant to be acceptable. Later, in the post-Soviet situation, worrying about nuclear safety was indeed regarded as a luxury, not least because "energy poverty," that is, a lack of access to modern energy services, was a fact of life in many post-Soviet republics.[128] At the Ignalina nuclear power plant and in Visaginas, people feared the loss of employment much more than they feared a nuclear accident or the radioactive waste stored on site.[129] On the level of the Lithuanian government, the fear instead concerned the possibility of losing energy independence from Russia, and consequently some of the country's economic and political sovereignty. Thus, the social and economic challenges of the situation far superseded any fear of nuclear power.

Overall, the dream of a Soviet paradise as expressed in the "nuclear way of living" in Sniečkus/Visaginas was betrayed twice: first by the Soviet collapse and then by the agreement with the European Union to close the Ignalina plant. The longed-for national sovereignty of Lithuania, partly reliant on energy independence toward Russia, was also betrayed, as the fate of the Ignalina plant was regarded by many as too high a price to pay for the dubious good fortune of belonging to a new supranational union. The European Union was assuredly spending money on radioactive waste storage, but not contributing in any comprehensive way to solving the issue of energy supply in Lithuania.

Memory Work

How to deal with the rich and complicated past of the Ignalina nuclear power plant and the town of Sniečkus/Visaginas? There are many groups of people that could claim Ignalina's history as theirs. Most obvious are perhaps the highly skilled nuclear engineers at the plant, the Russian-speaking majority living in the town of Sniečkus/Visaginas, and the larger nuclear community of the former Soviet Union. But there were also the Lithuanian environmental movement during perestroika, which eventually led to the formation of an overt liberation struggle, the Western specialists involved in safety improvement measurements of Ignalina during the 1990s, and the populations in neighboring countries, not least across the Baltic Sea, who equated the Ignalina plant with Chernobyl and projected their nuclear fears toward it.

So is anybody claiming the Ignalina nuclear power plant to be an expression of their heritage? As far as my investigation shows, the answer is no, not really. It is true that there is a local historian writing a chronicle of the workers' town; the plant celebrated its twenty-fifth anniversary in 2008 with a book on its history and relationship with the local community, and since 1995 there has been an information center at the plant, which provides access to old photos and films as well as informing visitors about present activities within the decommissioning process. From a broader perspective, the character of mono-industrial towns has been described and the history of nuclear power in the Soviet Union has been dealt with, as has the aftermath of the Chernobyl accident, among other things. Still, the Ignalina nuclear power plant is activated in neither a local nor in a national or international contemporary context as a means for memory work and future orientation. Why is this? Is the scar too complicated and vulnerable to be articulated as valuable? Is it unclear whose scar it actually is? Or is it just that—as in Malmberget in chapter 2—the land of home and production has not turned into an appealing visual landscape to be detected by the tourist gaze?

Traditionally, heritage processes have been an issue for celebrating national unity; from this perspective the Ignalina nuclear power plant does not easily fit into a master narrative of Lithuania. According to Marija Drėmaitė, the general framing of Lithuanian history has long been that of a rural and agrarian country.[130] Although industrialization certainly did take place, mirroring the processes of European and later Soviet development, the scale was always comparatively small, such that social phenomena important elsewhere, like trade unions or workers movements, never became prominent. This means that groups that have been active in promoting industrial heritage in many other countries do not really exist in Lithuania. In addition, industry was an unfamiliar element in the public understanding of a Lithuanian national identity. The most far-reaching industrialization period—the Soviet era—is associated mostly with destroyed landscapes, pollution, low-quality products, and huge, ugly complexes.[131] These buildings and structures are simply not regarded as "heritageable" from a national point of view.[132]

As a consequence, until recently almost all of the industrial sites that have been listed and labeled "heritage" in Lithuania are picturesque nineteenth-century mills and old bridges. Simultaneously, another logic is applied to larger nineteenth-century industrial buildings located in cities where the general Western trend of reuse, as outlined in the introductory chapter, has been adopted.[133] Former breweries have been turned into shopping malls and former workshops serve as settings for restaurants or art galleries. The red brick buildings have become an asset in current architectural fashion, without much connection to their past function.[134]

Considering these circumstances, it is not surprising that my question about potential heritage qualities of the Ignalina plant was met with hearty laughs, both at the Department of Cultural Heritage in Vilnius and at the manager's office at the Ignalina plant.[135] The idea was a joke, but no one seemed offended. It was a joke for the heritage department because it was perceived as a frivolous suggestion that would imply spending far beyond any existing budget earmarked for Lithuanian heritage work. The nuclear power plant was not even on the list of priorities selected for documentation and preservation. It was a joke at the former manager Viktor Ševaldin's office at the Ignalina plant because such issues were never their business. "We never thought about it much," as he says, and laughs again.[136]

In the 1970s, during the Soviet period, some amateur historians in Lithuania tried to draw attention to what they termed "technical heritage."[137] They were not very successful, but one conceptual heir of this work is to be found in the Energy and Technology Museum in Vilnius, inaugurated in 2003 in a former power plant in the city center. One of the exhibits features a model of the Ignalina plant. The model was created in 1983 in connection with the commission of the first reactor and shown in an exhibition in Vilnius at the time. In the information center at the Ignalina plant, an almost identical model is displayed. Both models can be turned on to illustrate the nuclear power process in cross-section by light and motion in an eye-catching way. Industrial models were a common feature in the Soviet Union. They were used to impress foreign visitors with smoothly working Soviet industries, without any risk of the possible disturbances that could occur in real factories. They were also used to educate and inspire people to contribute to the creation of the communist society.[138] Would these models of the Ignalina nuclear power plant do as triggers for memory, along with photos and oral histories?

The heritage department in Vilnius is currently investigating modernistic buildings from the Soviet period. The issue of Soviet and

Russian heritage, now within the borders of Lithuania, seems to be an issue that the department is grappling with: "How to approach it? What to do? What is of value?...Nobody understands."[139] The deputy director Algimantas Degutis depicts an overwhelming task: "We have to describe what is valuable because times are changing very fast."[140] At the same time, he believes that heritage concerns should not hinder the process of development, that is, new infrastructure and new building construction, and in the background, the obligation to "hear politicians' voices."[141] In his view, anything done to recognize the heritage value of the Ignalina plant must be initiated by plant management, a notion almost as funny as the heritage question itself.

At the Ignalina plant, Ševaldin tells that even though it has not yet been decided how Ignalina will be decommissioned, he believes the best solution would be to cover the two reactors with concrete, and if so there would be two cubic forms left "like a monument to this place."[142] The decommissioning director Saulius Urbonavičius instead brings up the new storage buildings for the radioactive waste, which are currently under construction. He says that there are no plans to save anything for the special reason of memory or heritage, but the storage building: "This is the monument!...That is true. That is a fact."[143] The overall attitude nevertheless has to be characterized as resigned. To think about the future is "not a good thing" but certainly one has to accept the circumstances, because "what else is there to do?" even if they are now working to "kill their baby"—the Ignalina nuclear power plant. To the question as to whether anybody will miss the plant when it is gone, I get the answer: "It is difficult to say."[144] Maybe commemoration cannot occur until there is a past worthy of commemorating or, in other words, the wound has to be healed into a scar before it can be acknowledged?[145]

I suggest there is grief here, for which it is difficult to find words, and thus it is difficult to articulate and frame. Among other things, this lack of articulation and, by extension, lack of agency imply that a conscious heritage process cannot take shape, so that laughter may be the only possible response. The experience of betrayal naturally differs between ethnic Lithuanians and Russian-speakers, between plant employees and Lithuanian government representatives, but they all share the history of the Ignalina nuclear power plant.

Currently, the plan is to raze all structures (apart from a possible concrete covering of the two reactors). The process of demolishing buildings and dismantling equipment that are no longer needed has begun, but the more general demolition is planned to start in 2025 and finish in 2029.[146] According to the decommissioning director, Urbonavičius, there is no previous experience of tearing down an RBMK that they can rely on "if Chernobyl is not taken into account, because it is different."[147] In 2010, there were about nineteen hundred people working at the plant, compared to thirty-five hundred when it was in operation (or sixty-five hundred employed in total, also including those working in town).[148] The first wave of shrinking personnel was solved by retirement and voluntary departures, but the second step will be more "painful" says Urbonavičius.[149] At the same time, however, there are plans to build a new nuclear power plant at the same site and negotiations with potential strategic investors are under way. If a new nuclear power plant is constructed, Urbonavičius considers that would be "a better monument" since it would be a reminder of its predecessor for many years to come.

Concluding Remarks

The Ignalina nuclear power plant and the town of Sniečkus/Visaginas are primarily to be understood as land, that is, a place where people have their home and work. To the inhabitants in Lithuania not living in Visaginas, the plant has significant symbolic value indeed, connected to Soviet oppression and national energy independence. To many people in the West, Ignalina was one of those dangerous nuclear power plants of the Eastern bloc. In spite of the symbolic projections, I argue that this territory at the borders of Lithuania, Latvia, and Belarus has not turned into a landscape. It is neither part of a growing interest in Soviet nostalgia, nor part of the increasing tourism directed toward dark and dangerous sites. Former military prisons in Latvia attract large number of visitors, as does the severely contaminated exclusion zone around Chernobyl, Ukraine, since they are connected to experiences of thrill and ruination sought-after by wealthy Westerners. In contrast, the Ignalina plant represents a kind of "anti-landscape," a wasteland awaiting new investment that could bring hope to the community, or in due time to be concealed as a storage site for radioactive leftovers.[150]

The highly radioactive spent fuel, still kept in water basins in the reactor buildings, is a scar that could easily be turned open into a deadly wound. The risk of nuclear accidents has certainly been reduced since the plant was closed down, but the waste will remain dangerous for one hundred thousand years to come. The time perspective is breathtaking but not really considered at the Ignalina plant at the moment. An intermediate storage building is under construction and will last for fifty years. After that, it will be future's problem to deal with the waste. Somebody else has to find a way to convey the message of this scar—"Danger!"—to the people of the coming decades, centuries, and millennia.[151]

Whose scar is the Ignalina nuclear power plant? I have argued that there are many groups of people that could claim the plant's history as theirs, but nobody really does. The several overlapping significances connected to the plant stand out fairly clearly; what is lacking is an articulation and valuation of these significances as a whole, which, in turn, could reveal how competing and conflicting they might be. What is a legitimate politics of memory here, to rephrase Paul Ricœur?[152] My answer would be that this scarred landscape bears witness to two critical dreams of our recent past—the dream of the Soviet paradise and the dream of the independent nation-state of Lithuania having experienced occupation. Both dreams were carried by people who worked hard to reach them, and in some respect they even did. But the dreams were challenged and to some extent betrayed, and the continuing struggle, the ethnic and social tensions, and, the continuous economic and physical risks make it hard to articulate the traces of the lost utopia. Acknowledging the scar therefore becomes a question of luxury.

Chapter 5

Industrial Nature

Abandonment is often followed by overgrowing. When an industrial plant is abandoned, the growing plants bear witness to remaining substances in the ground—leftovers from industrial production. Following pioneering species, the industrial site will soon become marked by shrubs and later by trees, sometimes turning the place into a kind of forest. This "industrial nature" is acknowledged by botanists and ecologists as a habitat very rich in species. It is conceptualized by planners and landscape architects as an iconic post-industrial landscape, and has been turned into a visual genre by photographers and filmmakers epitomizing decay within a romantic or dystopian setting.[1]

The combination of industry and nature might seem dichotomous and value laden: the human and the nonhuman, the altered and the pristine, the exploiting force and the creating one, the black and the green. However, other understandings propose that industry and nature are intimately intertwined, acting as coconstitutors of new landscapes. In this way, industry and nature resemble other opposites that this book touches on, like industry versus culture, as expressed in resistance toward the idea of industrial heritage; or cultural heritage versus natural heritage, as expressed in the criteria for the World Heritage List—opposites that have moved toward increased entanglement in a similar manner. It has even been suggested that the concept of (nature's) "resilience" can be understood as a transformation of the concept of "cultural landscapes," which illustrates an ongoing scholarly effort to apply perspectives that emphasize the "hybrid and messy connections of the world"—without doubt a key feature of industrial nature—and also bring questions of responsibility to the fore.[2]

One post-industrial site where industrial nature plays a decisive role is the Landschaftspark Duisburg-Nord in the Ruhr district in

Germany. Here visitors encounter a former ironworks where spontaneous vegetation thrives in and around industrial structures and where strict rows of newly planted leafy trees occupy some open spaces in between. Many of the activities going on in the landscape park are transferred from a setting in "nature" as it is more commonly perceived, activities such as climbing, biking, diving, playing, and walking the dog. Yet the climbing paths follow the walls of former ore bunkers, the biking paths meander above and between the toxic ground of the former industrial area, and the diving takes place in a former gasometer. These and other happenings like concerts and art exhibitions have become so popular that the Landschaftspark Duisburg-Nord is said to be the most visited place in the seventeen-million-strong German state of North Rhine-Westphalia today, second only to the cathedral in Cologne.[3]

What is it that makes this post-industrial landscape so attractive? What makes it an object on the contemporary traveler's map of

Map 5.1 The Ruhr district is heavily industrialized and densely populated. The land is mostly flat, with few obvious reference points. During the 1990s, the former ironworks of Meiderich in northern Duisburg was turned into a "Landschaftspark," among other things offering new views over the environs. Map: Stig Söderlind, 2013.

spectacular experiences? The Western world has been grappling with redundant large-scale industrial areas for some decades now, and different strategies and outcomes can to some extent be evaluated and judged. Does the landscape park in Duisburg provide an optimistic answer to the overwhelming issue of what to do with brownfields all over the world? Is there a flipside to the shining coin—losers behind the scene whose voices are drowned out by the enthusiastic praises of brigades of visitors? The examination of this "post-industrial regime of nature" or "post-industrial urban ecologies" is still limited from a humanist scholarly perspective and, thus, the aim of this chapter is to explore the potentials of industrial nature within the framework of the landscape scar metaphor.[4] The empirical node is the Landschaftspark Duisburg-Nord in the heavily industrialized and densely populated Ruhr district in Germany—an area with a complicated history (map 5.1). A few detours take us to places like Berlin in Germany, Detroit in the United States, and Avesta in Sweden. What remedies are offered by industrial nature?

Industrial Nature and the Ruhr District

The physical landscape of the Ruhr district consists mainly of meadows and plains, bordered by the Ruhr river to the south, the Rhine to the west, and the Lippe river to the north. Coal—the basis for the regional industry and the dense settlement since the 1830s—is typically found in tilting layers in the ground. The coal layers come to the surface in the southern part of the district and go deeper and deeper further north. Hence the oldest remains of coal mining and early industrial activities are to be found in the south, close to the Ruhr. As mining techniques developed, the industry followed the coal layers further north; at the beginning of the twenty-first century the industry was concentrated just north of the small river Emscher, a tributary of the Rhine.

Coal mining and steel production in the Ruhr district were of decisive importance for German forces during both world wars and thus also an obvious target for allied occupation and bombing. Large parts of the Ruhr industry and built environment were destroyed during World War II and rebuilding was a given postwar priority. This effort was partly financed through the Marshall Plan and was an integral part of the European Coal and Steel Community created in 1952. The region that had been so crucial for the war was to become crucial for the peace, and the Ruhr industries also boomed in the 1950s.

Toward the end of the 1950s, coal mining was hit by crisis, due partly to overproduction and rising international competition, and partly to competition from new energy sources such as gas and oil.[5] A wave of closures followed. Just over a decade later, at the beginning of the 1970s, the oil crises occurred, severely affecting the Ruhr area. The almost 200 mines that were active at the beginning of the century dwindled to 125 in 1960 and to 29 in 1980.[6] Between 1960 and 1980 the number of workers in the mines shrank by 50 percent and, although the service sector increased in importance, it was not enough to replace the vanishing industrial jobs.[7] In the mid-1980s, the Ruhr had the highest unemployment rates in what was then West Germany, at over 15 percent, with large groups of immigrant workers among those most exposed to the situation.[8] The Ruhr district was described "as a giant dying from physical space and social issues diseases, besides the negative region image due to the soil contamination, huge unemployment and little creativity on the labour market."[9] How did industrial nature enter into this picture?

The origin of the concept of "industrial nature" is not entirely clear, but it appeared as a concept within botany in Berlin in the 1970s.[10] After World War II and up to German reunification in 1990, the special situation of Berlin as a divided city decisively influenced the emergence of urban-industrial nature as a prominent feature. In West Berlin large areas that had been destroyed during the war were not built up again, but left to their own devices for decades. Spontaneous vegetation flourished on these sites in successive stages of colonization, from herbaceous plants and shrubs to wild urban woodlands.[11] This was also the situation in several Berlin railway yards, since East Germany had been given the rights to control the railways and decided to reduce train transport in the western part of the city to a minimum. The plan of the West Berlin authorities was to spare unbuilt spaces as reserves for future planning, but also to provide green urban spaces to the walled-in inhabitants.[12]

The work of botanists and later ecologists eventually resulted in a conceptualization of urban-industrial nature as an ecosystem of "a fourth kind," as it was labeled by ecologist Ingo Kowarik, along with pristine, agricultural, and horticultural ecosystems. The nature of the "fourth kind" included industrial sites, but also more generally vacant lots and transport corridors.[13] Kowarik argued that these areas represented a new type of process that was important to acknowledge for many reasons, among them because the phenomenon was so widespread, because urban-industrial nature showed a special ecological configuration, and because the spatial location was within immediate reach of the public, which implied that urban-industrial nature could serve important social and ecological functions.[14]

The ideas from Berlin were thus to some extent established when leading politicians and business executives in the Ruhr district realized that the negative situation needed to be dealt with. A new image had to be established and the material legacies from the past had to be transformed in a way that could bring hope for the future. On a federal level as well, there was a will to invest in the Ruhr—the area being considered as one of the weakest parts of West Germany.[15]

In 1989, a ten-year program, the Internationale Bauausstellung (IBA) Emscher Park, was launched, inspired by the German tradition of building exhibitions. The aim was articulated as to promote urban development from a social, cultural, and ecological perspective.[16] From the beginning of the twentieth century the Emscher river, after which the IBA was named, had been an open sewer because ground conditions precluded the construction of underground sewers; to many, the stinking trench epitomized the gloomy image of the Ruhr district.[17] By this time the land in the immediate proximity of the Emscher had largely served its use and was said to be the area with the largest problem density in the country.[18] In addition, the regional organization of the Ruhr was a complex network of economic and political authorities, associations, and corporations, which did not really share a common vision and had difficulties cooperating.[19]

The program began with a call, sent out by the Ministry of Urban Development, Housing and Transport in the North Rhine-Westphalia, formulated by Social Democratic politician Christoph Zöpel and geographer Karl Ganser, soliciting project proposals from all sectors of society. A public corporation was set up to act as coordinator and to monitor the quality of the proposals, and later to facilitate realization of the ideas. The decision to let the overall program grow out of a number of individual projects had the benefit of allowing different actors to express their own ideals and emphases while at the same time contributing to a joint effort. The basis for the IBA program—which had to be integrated into all of the individual proposals—was an 80-kilometer-long landscape park that was to take shape along the Emscher river.[20]

With German reunification in 1990, the Ruhr district was suddenly no longer the weakest part of the country. The available funds for the IBA program became more limited and a new vision of "Change without Growth" was added to the transformation work.[21] At this time, there were more than eight thousand hectares of abandoned industrial land in the Ruhr district.[22] These vast areas could not easily be transformed into the well-known types of spaces for previous types of uses, such as new industrial production, residential areas, or well-managed parks. Along with the problems of decontamination, both actual demand and financial means were lacking. Thus, a new conceptualization of the abandoned industrial land was needed, and three central aspects of the IBA program became *industrial nature*, *industrial heritage*, and *industrial art*. Through this focus, the existing environment was turned into a material and imaginative asset that was activated in an ambition to change the future course of the district.

Abandonment, Discovery, and Spectacle

One of more than one hundred individual projects carried out within the IBA program was a transformation of the ironworks Meidericher Hütte in Duisburg into the leisure area Landschaftspark Duisburg-Nord.

When the ironworks in Duisburg, in the western part of the Ruhr district, closed down in 1985 after more than eighty years of operation, no one lost their job, as workplaces simply moved to a more modern plant nearby. In spite of the generally negative situation in the Ruhr, this meant there was no feeling of crisis directly connected with this closure. The owner, Thyssen AG, kept and maintained the Meidericher Hütte for about a year, to keep it in reserve until all production had been successfully transferred to the new plant. After this period in limbo, it was generally expected that the company would demolish the old ironworks.[23] However, instead a group of local citizens began to argue that the former ironworks should be preserved, and eventually they succeeded in their efforts.[24] A public company acted as trustee for the city of Duisburg and became the new owner, buying the 230-hectare ironworks area on the outskirts of the city from Thyssen for a symbolic sum.[25] For Thyssen, this solution was decisively cheaper than keeping and decontaminating the site.

The transformation of the Meidericher Hütte thus began before the IBA program was launched, and the planning process leading to the IBA Emscher Park was influenced by what was going on in Duisburg. According to Wolfgang Ebert, then president of the German Society for Industrial History and involved in the work with the Meidericher Hütte, the former ironworks constituted a core and "a small galaxy of the overall ideas" that later characterized the IBA Emscher Park.[26]

Figure 5.1 When the former ironworks in Duisburg opened as a landscape park, visitors could climb one of the blast furnaces and get an extensive view, both of the post-industrial site and of the surrounding landscape. Photo: Anna Storm, 2007.

Therefore, to some extent, it became a test arena for the larger-scale program. What, then, marked the transformation of the ironworks into a landscape park?

Having secured the preservation of the ironworks, the earlier "forbidden city" was gradually opened up to the public. Special guided tours were offered, and early on it was possible to climb up a blast furnace via a prepared stairway to enjoy an extensive view over the landscape (figure 5.1). For most of the visitors it was, Wolfgang Ebert recalls, a "surprising sensation to go in there, for they had never been able to see something like that; despite the fact that they are surrounded by this landscape, they do not know what is in there."[27] This story of discovery has its counterparts in many post-industrial landscapes, pinpointing the fact that the transformation attracted mainly not former workers, but people without previous direct experience of the ironworks. At many sites it was newcomers rather than locals who had first emphasized the potential for new uses at the abandoned industrial sites.[28] In the case of the Meidericher Hütte, the astonished gazes of the first visitors were directed both toward the formerly hidden place of the ironworks and toward the overall industrial landscape of the area. Since the flat land of the Ruhr was almost completely lacking in vantage points, the possibility to see afar from the top of a blast furnace was a great part of the attraction.

The IBA program launched a landscape architecture competition for the Meidericher Hütte. The winner was the firm Latz und Partner, and from the early 1990s they worked on the transformation process. The former industrial area—dominated by the abandoned blast furnace plant, casting machines, railway lines, overhead cranes, ore bunkers, storage facilities, and administration buildings—was also given a new name: Landschaftspark Duisburg-Nord. In comparison to other suggested names, such as "Feuerland" ("Land of fire"), the choice of "landscape park" sent clear signals of a move from industrial toward natural allusions.[29]

Accordingly, the emerging industrial nature at the landscape park was recognized early on. At a first stage, so-called ruderal species began to colonize the empty and silent industrial structures. "Ruderal" is a term describing both a type of ground and the vegetation growing there. The word has its origin in Latin, "ruderalis" and "rudus" meaning rubble, like gravel and broken bricks, and typical locations for ruderal species are roadsides, ruins, and other leftover or in-between spaces. The word is both an adjective and a noun, used in plant science from the eighteenth century to identify habitats highly disturbed by human activity and, from the late 1970s on, also as part of an ecological classification system called the CSR triangle theory.[30] More generally, the concept is used to describe any wasteland, such that "ruderal species" are sometimes equated with weeds. Ruderals are usually pioneers, fast-growing plants that rapidly complete their life cycles and often produce great numbers of seeds. They typically dominate a disturbed area for a few years before gradually losing out in competition with other species.[31]

At the Landschaftspark Duisburg-Nord, this first layer of spontaneous vegetation was complemented by conscious planting. Leafy trees that blossomed with light pink flowers in the springtime were embedded in strict rows between the industrial structures. The combination of low growing ruderal species and slender trees within an industrial setting was introduced to the public in lavish photo books and through guided tours with "industrial nature" as their special focus. In these presentations the abandoned industrial structures were acknowledged for their former function, but the main emphasis was put on their dramatic sculptural qualities in relation to the growing vegetation. The industrial and the natural elements were thus not regarded as contradictory, but instead as mutually reinforcing in the landscape park.

The landscape architect Peter Latz envisaged that "in time, the greenery will dominate the technical constructions of the gateways. So bit by bit another history, another understanding of the contaminated site and of the idea of the 'garden' is developing."[32] In keeping with this prophecy, after some time the spectacle of contrasting industry and nature began to include a friendlier everyday experience as well, and it was asserted that "the fear of pollution and contamination has given way to a calm acknowledgement of the old structures."[33] Moving away from an unambiguously industrial area toward the conception of a landscape park and the "idea of a 'garden'" might seem to be a long journey upon the same piece of land, but as we will see later in this chapter, it is a journey repeated elsewhere in the Ruhr district.

However, the Landschaftspark Duisburg-Nord was characterized not only by an appreciated process of overgrowing, but also by spectacular sports and cultural activities. As mentioned at the beginning of this chapter, a former gasometer was turned into a diving facility, open to the public and hosted by the Park Diving Club, while former ore bunkers provided walls for a number of climbing paths, similarly open to the public and much used by the local division of the German Alpine Club (figure 5.2).[34] Both outdoor and indoor spaces of the

Figure 5.2 Among many new activities in the landscape park, the local division of the German Alpine Club has constructed climbing paths in the former ore bunkers. Photo: DAV Duisburg.

former ironworks were adapted to house concerts, art, history exhibitions, and marketing events. There were outdoor cinema shows and a specially designed light installation in the evenings. During the ten years of the IBA Emscher Park program the park was also the location for nearly sixty movie productions.

A similar translation of activities usually taking place in "nature" took place in Bottrop, not far from Duisburg. Here an indoor ski slope, built as a pipe, wound over six hundred meters down a slag heap, and the possibility of downhill skiing in Bottrop was much appreciated both by people living in the vicinity and by people traveling longer distances, for example, from the Netherlands.[35] All year round you could ride the escalator to the top of the waste heap, enter the −4 °C atmosphere and go skiing, or you could sit in the Alpine chalet at the top and have a beer while listening to music from Tyrol.[36]

Certainly it was discussed what uses of the former industrial area in Duisburg were considered appropriate in its new shape as a landscape park. This was a debate that resonated with the larger IBA program and also beyond in later transformation projects in the Ruhr. A key question seems to have been how to direct the transformation process in a way that resulted in "site specific" characteristics and "authentic" places.[37] Historic, aesthetic, and ecological aspects were regarded to be crucial in this respect or, expressed in another way, "that interpretations and activities should evolve from their own environment."[38] Wolfgang Ebert, for example, did not consider performances of operas or rock concerts with a blast furnace in the background at Landschaftspark Duisburg-Nord to be "authentic" events, but mere translocations of rather ordinary events. He argued that the trend of using industrial sites as stages was dangerous, because it could soon fall out of fashion and then suddenly nothing would be left.[39]

Sociologist John Urry has highlighted how our experience of places has become increasingly governed by visuality, which has in turn developed into an "abstracted, disembodied quality or capacity."[40] He argues that all places therefore tend to become more and more cosmopolitan and nomadic, that is, to lose their unique meanings. The sought-after site specificity and authenticity at Landschaftspark Duisburg-Nord and within the IBA program at large can probably be understood as a response to this tendency.

The IBA program was much applauded but also criticized, among other things for being elitist and for not taking enough consideration of local populations and communities.[41] It was argued that the Ruhr was "reconstructed for a not-yet-existing middle class," explaining the low involvement of the local working class population in the process.[42] The transformation work at the Landschaftspark Duisburg-Nord showed a typical mixture of almost no former workers involved at an initial stage and, later on, a group of retired workers and engineers voluntarily engaged in maintenance work. None of the projects in the IBA program focused primarily on new work opportunities, but instead on transforming the land along the Emscher river into places where people would feel at home. According to a study by geographers Andreas Keil and Sibylle Findel, important steps in this direction had been taken by the end of the IBA.[43] Here we can notice an explicit combination of the features of "land" and "landscape" as outlined in the introductory chapter, a combination of the places of "home" and the places of "visual desire."

At the beginning of the 1990s, most of the industrial remnants in the Ruhr district were generally perceived by local residents as abandoned, polluted areas and symbols of economic decline. A decade later, this image had changed so that the industrial structures were viewed more as "relics of culture," with people living close to transformed post-industrial landscapes using them actively and identifying them as something positive.[44] This was true for children and teenagers as well as adults, and industrial nature was a key ingredient of

Figure 5.3 Traces of young lovers at the top of a blast furnace in the former ironworks, indicating its transformation into a place for adventure and refuge. Photo: Anna Storm, 2007.

the positive interpretation since it provided sought-after places for unregulated play and contemplation (figure 5.3).[45]

As one part of the study by Keil and Findel, people using the Landschaftspark Duisburg-Nord were asked to take photographs, and a frequent theme chosen was the contrast between "wild" vegetation and old industrial structures.[46] The photographers commented that they were "impressed especially by the power of nature and the frailness of human production" and that this motif gave them "hope for the future."[47] A resident in the vicinity of the landscape park responded to the investigators as follows:

> How I see the area? Like a sort of paradise garden. Because all the time something new appears, you can observe something different growing at every time, and the colours, it is an enormous, gorgeous variety. That is the fascinating thing about it. The transitions of the seasons in spring and the autumn are pure, that is, they are totally extreme and wonderful.[48]

It is not entirely easy to decide whether these visitors or users of the Landschaftspark Duisburg-Nord found the place attractive because of the entanglement of industry and nature, or if their positive response was based mostly on the growing vegetation, *in spite* of the existence of the industrial structures. In any case, to many the landscape park had become a place to experience change in a way that included the intimate conception of a paradise garden, excluded from the surrounding world but still easily accessible and attractive all year round.

When the IBA program finished in 1999, the Landschaftspark Duisburg-Nord played a central role in the festivities, among other things, housing the final exhibition that presented the whole program.[49] At the time, 65 million D-marks of public money had been spent on transforming the Meidericher Hütte into a landscape park—a transformation that was a way of dealing with the past in a perspective of the future, characterized by discovery, spectacle, and continuous change.[50] Now, let us take a step back in time and explore a more explicit heritage perspective in relation to industrial nature in the Ruhr.

Heritage and Nature in Coexistence or in Conflict?

As in many other industrialized Western countries, the 1960s brought different trends of recognizing industrial heritage in new ways to the Ruhr area, closely related to the ongoing structural changes. The main perspectives were, on the one hand, an emphasis on architecture and aesthetics and, on the other, an articulation of social "history from below." Both approaches were to some extent headed by an intellectual elite, although individuals usually focused on one of the two approaches rather than combining them.

One industrial site that attracted interest early on, mainly from the perspective of aesthetics, was the Zeche Zollern II/IV, a coal mine from the turn of the twentieth century in Dortmund. Some of its overground buildings were built in Art Nouveau style, designed by renowned architects and thus easily put into an architectural preservation discourse. When the mine was shut down in 1966, artists and intellectuals prevented its planned demolition and contributed to its becoming the first protected industrial building in West Germany in 1968.[51] During the following decades, industrial built environments became canonized as heritage in many parts of the country, with special funds set aside in North Rhine-Westphalia from 1975 and onward for the preservation of industrial monuments.[52] Notable events confirming this domestic articulation of industrial heritage values are the UNESCO designation of the Völklingen ironworks in Saarland and the Zollverein coal mine in Essen as World Heritage Sites in 1994 and 2001, respectively.

The "history from below" or "*Geschichte von unten*" began in West Germany as a movement to preserve workers' housing that was threatened by demolition during the 1960s. In the 1970s, the groups of intellectuals were joined by people from the working class; protests like spontaneous squatting, for example, became a common way to combat the demolition of both residential and industrial buildings.[53] Furthermore, and again similar to many other industrial areas in Western Europe, West Germany saw a new type of museum appearing: decentralized industrial museums with a social history agenda and monuments left in situ, in combination with museum exhibitions located in former industrial buildings.[54] The number of newly established museums or exhibitions focusing industrial history and heritage continued to grow during the 1980s and 1990s, resulting in a museum landscape in North Rhine-Westphalia comprising a couple of larger museums and more than one hundred smaller museums dedicated to industrial history.[55] Somewhat contradictory to this convincing picture that the experience of industrial work was being preserved, Dagmar Kift at the LWL-Industriemuseum in Dortmund rhetorically suggests that while labor first disappeared from the mines, steel works, and factories into the museums, it now seemed to disappear as a key topic *within* the museums. However, she concludes that this is not a result of labor being uninteresting to people, but because it needs to be brought into line with a broader range of relevant issues such as the perspectives of migrant and female workers.[56]

So how does an appreciation of industrial architecture and an articulation of "history from below" or labor history, on the one hand, relate to industrial nature on the other? I argue that it is not one single relationship, but a range of different entanglements, as will become apparent in the following section, in which we explore a few examples from other places.

One of the aspects where labor history encounters industrial nature concerns the way nature has been used as a way to conceal social justice. In a case study of the contemporary conversion of former military arsenals and nuclear facilities in the United States into nature refuges, geographer Shiloh Krupar shows one way in which industrial territories are transformed into new landscapes. In the transformation, the experiences of former workers, including radiation exposure and the politics of secrecy, were played down and hidden behind rhetoric concerning the technical monitoring of waste and the establishment of nature protection zones. The result was that the legacies of the past were neglected or simply denied, and that political responsibility became located elsewhere, beyond the reach of those concerned.[57] Krupar asserts that spectacle and uncertainty were central power techniques in blurring the previous sacrifice zones of nuclear industry with the new protection zones connected to nature into a postmilitary nature refuge.[58]

A similar perspective is proposed by geographer Nate Millington in his work on photographic depictions of declining, overgrown Detroit, Michigan, in the United States. Millington argues that a special genre or visual story has been established during the last decades, in which the growing plants are seen as either destructive or redemptive, but always as separate from the human-built city. The main motif in the photographs is the contrast between decaying buildings and growing vegetation; seldom do any people appear in these images. The strength of this visual genre contributes to hiding the actual ongoing lives of the city, and the decline is depicted as inevitable and apolitical, which, of course, it is not. Instead of justifying the present situation, he argues, the understanding of Detroit must include alternative futures.[59]

Yet another facet of concealing social justice by adding nature is framed within the more generally encompassing concept of gentrification. There are numerous examples of how a former industrial place was reused and gentrified, partly within an ecological framework, with the result that previous inhabitants were forced to move away, and previous industrial activities with a negative environmental impact were moved to other countries that could not afford to concern themselves with pollution.[60] This kind of transformation of redundant industrial sites often highlights conceptualizations of a desired urbanity, partly composed by natural elements like waterfront location and softening greenery in a "staging of urban nature."[61] Geographer Chris Hagerman notes how the concept of "livability" is based on a "complex and unstable set of understandings combining ideologies of nature, society, urbanity and nostalgia," pinpointing contemporary ideals of urban environments that include both vegetation and references to the past.[62] In his example from Portland, Oregon, in the United States, experiences of industrial production were played down in favor of a nostalgic depiction of a pre-industrial situation. Difficult pasts of a more recent industrial period were avoided, while other industrial aspects were turned into commodities. The result became new "landscapes of consumption" from which legacies of oppression and pollution and the history of workers and residents was largely omitted.[63]

Allowed overgrowing, conscious planting, and wildlife management in industrial or urban settings as analyzed by Krupar, Millington, and Hagerman have thus become expressions of visual rhetoric, concealing difficult experiences of local residents, at the same time as they contribute to avoiding politically controversial topics concerning contemporary responsibilities.

A rather different debate is to be found in connection with a former ironworks in Avesta, Sweden, where both spontaneous and planted vegetation were linked with architectural qualities, historic authenticity, and future use of the area. In a transformation of the ironworks area into an accessible part of the town center, favoring cultural institutions and small-scale service businesses, the issue of cleaning and greenery was a topic of diverging views in the late 1990s and early 2000s. On the one hand, most local opinion thought that the area should be cleaned and the visual impression softened by new greenery. This view was also shared by municipal representatives for the business sector, eager to create an attractive spot for companies to establish their offices.

A contradicting perspective was proposed by antiquarians and representatives of cultural institutions. Their argument was that vegetation destroyed an authentic appearance of the rough and rusty industrial environment and diminished its value, both from a heritage perspective and from a more general cultural perspective. Eventually, a compromise was reached in which the area was changed by some planting, but spontaneous vegetation in direct connection to industrial buildings was removed.[64] Nature was treated as both an asset and a threat, but overall as something that had to be controlled and given shape, perhaps more in line with the horticultural nature of the "third kind," like street trees, well-managed parks, and gardens, than corresponding to the urban-industrial nature of the "fourth kind" described above.

These examples show both similarities and differences in comparison to the understanding of industrial nature in the German Ruhr district. The Swedish and the German strategies share, for example, an overall focus on urban renewal through cultural and economic development, while the US examples are more inclined to focus on issues of public health or social justice. This difference has been interpreted by landscape architects Wolfram Höfer and Vera Vicenzotti as a general divide between Europe and North America in their approaches toward post-industrial landscapes. They argue that even though post-industrial landscapes resemble each other over the world, the way they are perceived depends on cultural context and, essentially, on different concepts of landscape. In spite of significant similarities in physical prerequisites, the discourses and the management of the post-industrial sites differ considerably between the two continents.[65]

Höfer and Vicenzotti describe how "landscape" in a European setting is most commonly associated with homeland and a heritage that requires stewardship. In contrast, the US narrative of an immigrant nation includes a view on "landscape" as presumably virgin land that needs to be conquered.[66] In Germany, Höfer and Vicenzotti argue, industry is no longer interpreted as a destroyer of landscape but as a harmonious part of it, mainly because of a number of discursive traditions, among them an idealization of industry and a discovery of the nature specific to the former industrial sites.[67] Historians Thomas Lekan and Thomas Zeller further pinpoint how environmental debates in Germany have generally focused on the best way to harmonize cultural and natural landscapes, rather than asserting a sharp dichotomy between the two.[68]

The debate in the United States has been less focused on the cultural meaning of these sites, and more on their potential economic uses and the engineering challenges they present.[69] So while the aim in Germany was to foster economic development through urban renewal, the US focus was to improve public health through the reduction of possible exposure to contaminants.[70] However, in parallel to this depiction of difference in approaches, there was also mutual inspiration between individual post-industrial transformation efforts, for example, between the Sloss Furnaces in Birmingham, United States, and at the Meidericher Hütte in Duisburg.[71]

In sum, the encounter between heritage and nature in a post-industrial setting led to many different outcomes and interpretations. In this context nature was perceived as a way to conceal past and present social injustices, or as an element threatening the authenticity of a former industrial site or, on the contrary, as an element enhancing the experience of a former industrial site, turning it into a commodity in a gentrified "landscape of consumption."[72] In relation to this palette of predominantly pessimistic viewpoints, the activities in the Ruhr can be regarded both as extensions of trends visible in other places, and as unique in principle. The optimistic view on resilience as a transformation of the concept of cultural landscapes, mentioned at the beginning of this chapter, is perhaps most fully expressed by the ideas marking the changes in the Ruhr, and less discernible in the other cases. Let us take a closer look at this entanglement by considering not what industrial nature might hide, but what it could make visible anew.

An Alternative Beauty of the Post-Industrial Landscape?

The examples described so far have shown industrial lands that were turned into post-industrial landscapes, in terms of appearance and through gentrification, often understood as something primarily negative and delimiting as compared to the treatment of past experiences. Is there an alternative view on the post-industrial landscapes in the Ruhr context, defined by a conscious integration of industrial nature into the cultural landscape setting? Is there another post-industrial beauty to be imagined?

The central position of visuality in the contemporary human experiences of landscape has been framed in many different ways, for example, as a choreographed "way of seeing," being more important than bodily experience.[73] The tourists' world, says sociologist Zygmunt Bauman, is "fully and exclusively structured by *aesthetic* criteria," such that artworks are highlighted as a means to see places with new eyes and as tools in the process of de-industrialization.[74] What if we apply these statements to the transformation work carried out at the Landschaftspark Duisburg-Nord and within the IBA program more generally?

Making landscape visible by means of artistic interpretations or commentaries was an approach much used in the IBA program in the Ruhr. The appreciated and indeed new outlook from the top of a blast furnace in the Landschaftspark Duisburg-Nord had its counterparts in numerous pieces of art placed, perhaps not on blast furnaces, but on top of slag heaps all over the Ruhr district. These aboveground storage sites for waste materials from the mining industry made up sought-after hills in the flat land, and also became nodes of orientation when turned into venues for art installations. Most of the installations in the Ruhr took their point of departure in the industrial character of the area, and the materials used were usually steel and concrete. And, I argue, even though the industrial structures of the former Meidericher Hütte in Duisburg are not artistic installations per se, a post-industrial gaze successively turned them into a kind of sculpture. Thus, the new layers of the industrial land of the Ruhr— the industrial nature in combination with pieces of art—contributed to make it visible anew, no longer as industrial land, but as a post-industrial landscape.

Another way in which the landscape has become visible is through the appreciation of ruins. Lately there have even been international debates concerning the pros and cons of "ruin porn," that is, photographic depictions of decay. In many places industrial ruins have turned into hiding places and sites for fantasy and adventure, used by children and youngsters, criminals and addicts, lovers and scrap dealers.[75] Human geographer Tim Edensor asserts that society has become so overwhelmingly organized that people are trapped in the predictable; in this context, the ruins offer a rare place of disorder and openness for interpretation.[76] Another tempting aspect of ruins is an experience of authenticity, of time passing by or, as phrased in an early memorandum of the IBA program, the fact that some abandoned industrial sites were selected for preservation to serve as "marks of truth from the past in the form of ruins."[77]

Previously, we saw how the issue of authenticity was debated regarding which activities were deemed to be appropriate at Landschaftspark Duisburg-Nord. We also saw how growing vegetation in Avesta was considered a threat to the authentic rusty character of the former ironworks. How could the concept of authenticity be further linked to industrial nature?

In the field of heritage, "authenticity" is mainly connected to cultural heritage in its tangible sense, and the equivalence for natural heritage is "integrity." In the valuation criteria of the World Heritage List, "authenticity" refers to places being original in form, material, design, and location, while "integrity" refers to places being typical and with interdependent components. Both are expressions for a modern search for the untouched, the genuine, and the traditional, and, importantly, often turned into aesthetic qualities—implying that the most genuine place is also the most beautiful.[78]

From the perspective of ecology, too, there is a favoring of "nature" that is regarded "original" or genuine in the sense of being most "natural." Ingo Kowarik suggests that there are two scientific approaches to naturalness: one retrospective and one prospective perspective.[79] The retrospective perspective traditionally dominates as it favors conservation of pristine ecosystems, that is, an idealized picture of an "original" natural landscape. The prospective perspective is less widespread and focuses instead on a natural capacity for process.[80] However, when the prospective perspective is applied, one may reach the perhaps surprising conclusion that among ecosystems in a city, the industrial nature, or urban-industrial woodlands as Kowarik calls them, are the most natural ones.[81]

Besides distinguishing between these two scientific approaches, Kowarik also proposes four types of nature, mentioned previously, as a way to conceptualize and differentiate urban green for people not so familiar with ecology and botany. The first kind of nature is the "original," that is, remnants of pristine ecosystems. The second is land transformed by agriculture and forestry, and the third is greenery that emerged through horticultural planting and maintenance, for example, gardens and street trees. The fourth kind of nature, finally, is the natural development that occurs spontaneously on urban-industrial sites, especially abandoned areas. Based on this conceptualization, Kowarik pinpoints how the second and third types of nature are very much culturally influenced, while the first and the fourth are defined more by the effects of natural processes, and thus the most closely associated with a wilderness character. To keep them separate, he suggests we talk about an "old wilderness" connected to the first, pristine, nature, and a "new wilderness" connected to the fourth kind of nature, the urban-industrial woodlands (figure 5.4).[82]

Thus, post-industrial landscapes with their industrial nature can indeed be understood as natural from an ecological point of view, even as a new wilderness, and to a great extent this new wilderness has become appreciated, at least in the German context. At the same time, from a traditional standpoint, overgrown post-industrial landscapes challenge heritage views on the original, authentic, and beautiful, as well as ecological views on conservation of the pristine "old wilderness." Let us now explore what the challenging, new

Figure 5.4 Spontaneous overgrowing as well as regular planting have become conscious strategies at many former industrial sites in the Ruhr district. One goal is to create attractive environments of a "new wilderness." The picture is from Landschaftspark Duisburg-Nord. Photo: Ralf Schmidt, 2009.

wilderness could comprise, both in terms of symbolic labeling and actual vegetation.

The Landschaftspark Duisburg-Nord was obviously framed as a park, among other things featuring industrial nature. Other sites in Germany were also framed as parks or as forests or gardens, and we will now touch briefly on some of these sites in order to put the former ironworks in Duisburg in a regional empirical perspective.

One place where the "new wilderness" of industrial nature was both favored and tamed within the idea of a park was the Schöneberger Südgelände, a railway yard in Berlin, left abandoned after World War II and successively overgrown. In the 1980s, local citizens fought to preserve it as a natural area, and 20 years later it became one of the first official conservation areas in Germany with protected, species-rich urban-industrial nature.[83] The goal was to keep the former railway yard open and accessible to the public and, at the same time, protect the rich flora and fauna that had developed during the decades of abandonment. In addition, it was acknowledged that the richest habitats existed in the industrial nature of early or intermediate succession stages, with biodiversity diminishing once vegetation reached the stage of actual woodland.[84] The solution from an ecological point of view was a mixture of free and controlled access for the public, and maintenance and nonmaintenance of the vegetation in particular zones. So, the park was managed to meet the needs of both humans and animals and plants, sometimes through separation and sometimes through combination.

Moving from the park to the idea of industrial forests, a large-scale experiment from the late 1990s called the *Industriewald Ruhrgebiet* ("Industrial Forests of the Ruhr district") is striking. Landscape architect Jörg Dettmar describes it as a possible strategy for handling the huge areas of abandoned industrial land in a way that does not incur heavy costs, and does not require much maintenance. One element was the legal and administrative aspects. Federal and state forest regulations allow for an area to be defined as a forest, even if there are only very early stages of vegetation and no "dominant woody layer."[85] This classification had significant advantages for the owners, since the standards for liability were lower than, for example, for a park.[86] It implied that abandoned industrial sites could be defined as "forests" with a forester taking care of them, even though there was not yet a natural environment that most people would describe as a forest.[87]

Finally, some industrial nature has also been termed and understood as "gardens," suggesting yet another set of allusions. A garden is small, intimate, and friendly, quite far from spontaneous imaginations of industrial land or forests. However, gardens in former industrial sites have emerged in the Ruhr and in several other places in Germany. The former Hansa coking plant in Dortmund, for example, is described as "a garden of abandonment," while there are twelve thematic gardens called "the Paradise" to be found at the Völklingen ironworks, opened in the midst of the former industrial area in 2009.[88] Within the well-established narrative of the discovery of a hidden and abandoned place, it is framed as a "mysterious garden of Eden" and an "island of wilderness," thus combining the ideas of intimacy with ideas of the wild. The change of seasons is said to drench "the paradise in vibrant colour and pleasant smells"—indeed something different for the senses if we remember the stinking open sewer of the Emscher river in the 1980s, an expression of the combination of industry and nature at that time. By alluding to the most pristine of all natures, Eden, a decisive break with the industrial past is made. The fourth nature is thus transformed—in an imaginative way—into the first, pristine nature, and the "new wilderness" in some respect has become the "old wilderness," yet one that includes a human presence.

In sum, I suggest there is an alternative beauty of the overgrowing post-industrial landscape articulated in the Ruhr district. At least, it is clear that these landscapes have become popular, and not only in Duisburg, where the wide range of attractions might make it tricky to substantiate such claims. The Natur-Park Südgelände, for example, is estimated to attract fifty thousand or more visitors per year; within the Industriewald Ruhrgebiet project, regular guided tours through the "industrial nature" have doubled their number of participants in a just few years.[89] Whether these visitors see the new urban industrial nature as a redemptive paradise or as reflecting a dystopian vision, the attraction of discovery, contemplation, and play seem to be comprehensive in our time.

A Possible Substitution Story

How should we understand the overwhelming focus on nature and landscaping in the restructuring of the Ruhr district? One possible track is rooted in the many difficult events of the past connected to the Ruhr district: World War I, the French occupation of the Ruhr district in the early 1920s, and World War II, the last perhaps of a particular indignity due to the question of guilt.[90] What is more, industrial crises, severe contamination of land, and extensive immigration have put tension on the people of the Ruhr even in the postwar period. How would it be possible to deal with the overwhelming material legacy of this complex past?

The shaping of history and heritage entails many choices about which aspects to stress and which to suppress. Newly independent nations, for example, often choose to sweep away a hated past and replace it with an older heritage.[91] Geographer David Lowenthal suggested that while individual forgetting is largely involuntary, collective oblivion is mainly regulated and has an aim.[92]

But are the stories of war and Nazism really suppressed, or perhaps concealed, in the Ruhr district? And if so, is there a collective purpose to be disclosed? As we have seen, there are certainly museums and heritage sites in the Ruhr dealing with the region's twentieth-century history. Nevertheless, perhaps the determined focus on the natural environment—industrial nature—presents a possibility, if not to replace, then to balance these more difficult elements of the past. If polluted nature could be seen as the main representative of the difficult past that contemporary society has to deal with, the focus of the IBA Emscher Park program on a sustainable recovery of nature could become more than a way of making the district a physically healthy place to live in. The work to overcome the environmental disasters of the past could then also be seen as a choice to deal with, in the words of sociologist Sharon Zukin, "a more manageable past."[93] In this respect, industrial nature as a substitution story can work as a scab upon the wounds of the past, or, from an opposite position, as a new meaning that obscures certain stories.

In his study on American landscapes of violence and tragedy, geographer Kenneth E. Foote argues that stigmatized sites in the United States often remain "scarred" forever, since there is a lack of rituals of symbolic cleansing that could allow their reinclusion in everyday life.[94] In a German context, and somewhat more mildly, it was noted how the perception of new, post-industrial nature is "damaged by the stigma of the painful social changes that made such nature possible."[95] In contrast, the IBA Emscher Park program director, Karl Ganser, stated that the Ruhr district in his vision should be neither "ugly" nor "beautiful," but it should be "good," thus articulating a potential for moving beyond the purely negative conceptualization of a scarred landscape, and at the same time recognizing the imperative of being "good" and not only visually pleasing.[96]

So what happened? Did the past become more manageable by acknowledging industrial nature in the post-industrial landscape of the Ruhr? Over the course of the decade-long IBA program, about 120 individual projects were carried out, most of them through the initiative of the public sector. The total investment was about five billion D-marks, one third of which came from private funding.[97] In contrast to this picture of heavy investment, there were nine active mines in the Ruhr district at the end of the period, as compared with 125 in the 1960s and 29 in 1980.[98] Transformation work continued when the IBA program was over, but many of the post-IBA projects were questioned in a new manner. For example, it was argued that the "kind of new development lacks the historical references and local sensitivities that made IBA projects so successful, replacing them with placeless, meaningless and cultureless landscapes."[99] Also, the interaction and participation of the local population was described as limited, so that social integration remained a challenge.

Other voices expressed more positive interpretations of the changes. We recall the study of user perceptions that showed that many people of different ages appreciated the industrial nature in the Ruhr as a place for refuge. On a more general level, the transformation process of the former industrial land has been described as a "creative tool for democratic and social progress, shaped as much by bottom-up initiatives as by top-down directives."[100] The changes have thus been seen both as a variant of the concealment of social justice, as analyzed by a number of US scholars, and in contrast, as a tool for democratization and empowerment. Paradoxically enough, the borders between these two diametrically different interpretations can be thin and diffuse. An activity that works to integrate in one respect might have the opposite effect from another perspective.

This thin border must be taken into account when considering the metaphor of a landscape scar in relation to the industrial nature of the Ruhr district. Mainly, the entanglement of industry and nature in the Ruhr was regarded as a way to create new, inclusive, and beautiful lands and landscapes where people could feel at home and travel as tourists. From this perspective—and by means of creating a substitution story—industrial nature could be described as a sign of healing the difficult past into a scar. A few critics instead saw industrial nature as one of the expressions of an elitist project that avoided issues of social and economic vulnerability. From this perspective, the painful wound of the past still existed, but was merely concealed by spectacular superficial cultural events and conceptualizations.

Consequently, the growing vegetation in former industrial sites in the Ruhr can be understood to both bring reconciliation and new hope and to obscure difficult pasts. The thin line between the two readings of the post-industrial landscape may be linked with the passage of time and associated with an emphasis on continuous change that marked the transformation process as well. Perhaps uniquely in relation to other former industrial sites in the Western world, actors in the Ruhr district explicitly addressed the issue of continuous change, thus challenging a common idea of desired stability. Change was articulated as a main characteristic of the landscape; the growing plants, the decaying buildings, the shifting perceptions of its residents and visitors. With a few exceptions, change was also valued as a resource. This has implications for the suggestion of a substitute narrative. If change is a main characteristic, then history is never finished, which brings a degree of liberation from past difficulties and questions of guilt, but also avoids the impression that "everything can be held on to."[101] For the visitors to and users of the industrial nature, the change was even part of the attraction, the shift of seasons and the constantly changing appearance of the vegetation.[102] Unsurprisingly, on the website for the Landschaftspark Duisburg-Nord, one can read that the park is "always in development."[103]

From the perspective of ecologists, change has been launched as a possible conservation strategy. The idea of "temporary conservation" suggests that urban areas should be set aside for about 15 years in order to enhance biodiversity. After this period, the area could be built on and used for industrial or residential purposes, while other areas should be set aside, or abandoned, for the sake of establishing species-rich habitats. If this strategy were applied, a spatial model of "mosaic cycles" of biodiversity would be developed, where different spots reflect different successive stages of spontaneous vegetation.[104]

The introductory chapter outlined how healing to form a scar was to be understood in the context of this book not as an automatic, linear process but as a cyclical, temporally uneven one marked by ambiguity. "Industrial nature" in the Ruhr district demonstrates such characteristics of the scar, and shows how the metaphor can encompass both the new, valued uses of the post-industrial landscapes and the elitist bias criticized for obscuring these landscapes.

Concluding Remarks

The conception of urban-industrial nature, a "fourth kind" of nature as developed from the work of botanists and ecologists in Cold War Berlin, has been activated in many different ways in the Ruhr and other industrialized parts of Germany, ranging from understandings of wilderness and actual forests to managed parks and gardens. Industrial nature can thus be seen as something that adds another layer of meaning to the changing industrial sites, in a palimpsestual way. Another approach, however, as was suggested in the introductory chapter, is to play down the retrospective memory of the palimpsest and instead imagine "a prospective memory, an unfolding and ongoing relationship between past, present and future" in an "increasingly fluid notion of temporality."[105] Seductively enough, a prospective memory resonates very well with the conception of a prospective approach to naturalness, in which the capacity for process is the key feature, rather than an imagined state of pristine origin.

Industrial nature challenges established categories or classifications, and has brought about confusion and insecurity. This confusion probably has to do with the spontaneous understanding of industry and nature as opposites, mirroring the distinction between the human and the ecological. Geographer Matthew Gandy suggests, however, that the growing presence of nature in former industrial landscapes "can be conceived as a kind of urban entropy" where these distinctions successively become blurred.[106] It is clear that many areas with redundant industrial structures have actually turned into an asset for a new understanding and appreciation of urban nature, facilitating a critique of the nature-culture divide embedded in the fabric of contemporary cities and modern planning.[107] Recalling the dystopian photographs of a decaying and overgrowing Detroit: while most images simultaneously mourned the city's current state and celebrated its picturesque aesthetic, some pictures actually raised questions about interrelationships and entanglements, about where the natural ends and the unnatural begins.[108]

Thus, in the search for ways to change the image and the physical reality of the Ruhr district the focus on the natural environment within the densely built industrial area became a key imaginary, that is, a vision connected to changed perceptions as well as to concrete action.[109] By using the concept of industrial nature, it was possible to combine the existing built infrastructure, along with its value dedicated to bringing historical identity, with strategic planting and controlled overgrowing as a means to achieve a changed experience of the Ruhr district. Industrial nature, as an aspect of the post-industrial landscape scar, thus constitutes a way to both conceal and to heal in the post-industrial situation; it focuses on a cyclical perspective, on temporal solutions for continuous change, but also pinpoints difficult issues of the inclusion and exclusion of past experiences.

Chapter 6
Enduring Spirit

Company towns are defined by the close connection between a company and a town. The company constitutes the only major employer, but the relation often extends much further. If the company is owned by an industrialist family, the family members are key figures in the town, and the company usually takes on extensive social obligations in the local community. A result could be mutual identification between the company and the workers in the town, evolving over decades. This model was prevalent not least in the Swedish company towns in the region of Bergslagen, called "bruk," up to at least the mid-twentieth century. This intimate coexistence marked the structures of the built environment, with the industrial area, the workers' housing area, and the industrialist family manor as central features.[1] The connection also influenced the way of doing things—sometimes expressed as the "spirit of a company town," or in Swedish, "bruksanda."[2]

A wide range of characteristics have been attributed to this spirit of a company town. Certainly it is an expression of power hierarchies between company management and the workers, including trust and loyalty. Furthermore the town forms a basis of a collective identification and one should not—as a worker or ordinary citizen—laud oneself.[3] What happens if this spirit is challenged by fundamental changes in the relationship between the company and the town? What happens in the local community when the family-owned industrial enterprise merges and becomes part of an international group of companies with foreign owners, when the family leaves the town, and when the production site is abandoned?

The slightly surprising answer to be found in the company town of Avesta, Sweden, is that the spirit seems to endure, transformed but quite unshaken. Here, municipal representatives have in many respects replaced the industrialist family as community leaders, so that loyalty is transferred to the town itself, based on strong ties to the past. New activities take place in the former industrial area, which was converted into a diversified cultural and commercial arena. Nevertheless, the process did hurt, and the town experienced several decades of uncertainty with a dilapidated industrial area in its midst, once the heart of the community, pumping blood to keep the system alive and occupying the central location between the town center and the river bank. So who and what were actually included or excluded in the choices of the transformation? Here is a post-industrial landscape scar to be examined.

A Place in the Forest, along the River

For centuries Avesta's economy has been based on forestry and the metal industry. In 1636, a copper works was established on a site along the Dalälven river. The ore was transported from the copper mine in the town of Falun, about seventy kilometers northwest of Avesta. Besides producing raw materials for household goods and roof plates, for many years all Swedish copper coins were manufactured in Avesta.[4] The main localization factor was the two falls in the river, which were used as a source of energy. The chosen site was also beneficial because products could be exported through the harbor in Västerås, and because of the surrounding woodland needed to make charcoal to fuel the smelting processes. At the end of the eighteenth century, the copper works was known internationally for its size and advanced technology, and had no counterpart in Sweden, and perhaps not even in Europe.[5]

However, as copper refining diminished in importance, in 1869 it was replaced by iron making. In 1883, the company Avesta Jernverks AB was founded. An ironworks, modern for its time, was established, performing all the parts of the process from ore to plate. Here were the blast furnace plant, an open-hearth plant, rolling mills, foundry, and mechanical workshop. The first blast furnace started operation in 1874, and stainless steel manufacturing began in 1924.

The copper works had been established on a long narrow piece of land along the south bank of the river, and the ironworks later took over and expanded within the same area (figure 6.1). At the end of the 1930s, almost all the land within the industrial area had been used. To ensure continued expansion the company planned to tear down a nearby workers' housing area. However, with the development of the electric grid, it instead became possible to establish a completely new industrial area, south of the town center and separated from the river.[6]

Figure 6.1 Avesta ironworks in 1912 with the Dalälven river to the right. Photo: Nordstjernan Corporation central archives, Engelsbergsarkivet.

The new area was called "the Southern Works" and the old area consequently started to be called "the Northern Works." To begin with, the company carried out production in both areas (map 6.1).

During the first decades of the twentieth century Avesta began to take shape as a modern town.[7] The Northern Works industrial area and the adjoining workers' housing area were separated from the new town center by a road, a railway, and a considerable difference in altitude, amounting to as much as fourteen meters.[8] When Avesta substantially expanded in the 1920s, the Northern Works was reduced to a district on the town fringe. The town expansion first took shape with relatively little planning, but quite soon initiatives launched to create a more urban built environment. A new town plan was elaborated in 1935, and during World War II the famous Finnish architect Alvar Aalto designed a new town center for Avesta, called "Acropolis."[9] The proposal included municipal and commercial localities assembled around the main square. It was Aalto's first design developed for a town center, and was probably his most important work in Sweden.[10] However, the design was regarded by local politicians as too spectacular and too expensive, so the issue was tabled.[11]

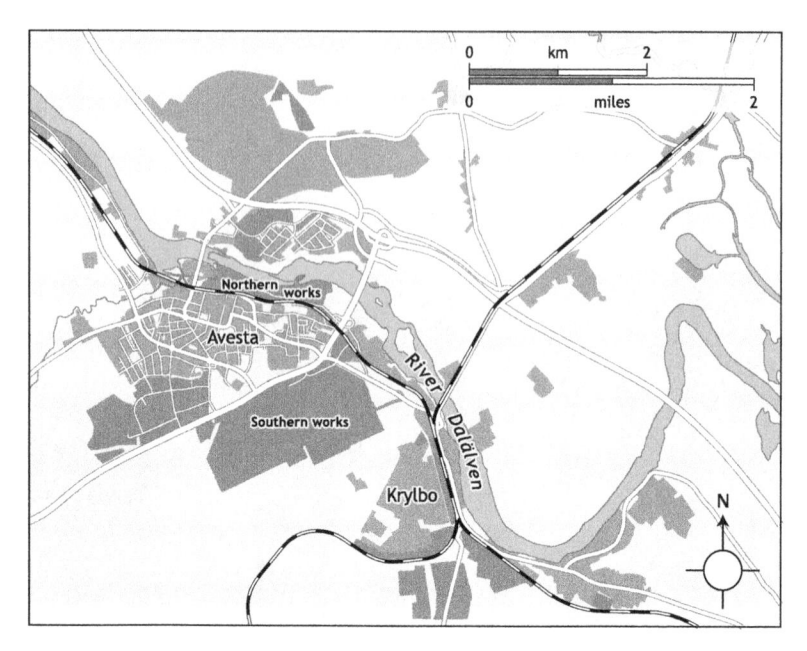

Map 6.1 The old industrial area of Avesta was located along the source of its energy, the Dalälven river. In the 1950s, the expanding production needed more space, so successively moved from the Northern Works to the newly built Southern Works on the other side of the town center. Map: Stig Söderlind, 2013.

At the end of World War II the ironworks employed approximately twenty-three hundred people; in the postwar decades the population of Avesta nearly tripled, from about seven thousand in 1950 to about nineteen thousand in the 1960s.[12] At this time—delayed because of the war—Avesta Jernverks AB began moving its production units from the Northern Works to concentrate on the Southern Works. The Northern Works, stretching almost one and a half kilometers from east to west along the river and about three hundred meters from north to south, was gradually abandoned, although the process was slow and affected different parts of the area in different phases. The western part, with the oldest and most spectacular buildings, like the old blast furnace plant built in slag stone brick and the open-hearth plant, both from the late nineteenth century, stood mostly empty. The central part, which contained a mixture of buildings from different periods, was partially used, but most of it was empty ground that had previously been used for storage, including land created by a filled in tributary of the river. The eastern part contained more modern industrial plants that were still used for production. Since some production continued at the Northern Works, the area remained fenced in and the two gates, one at each end, were kept closed and guarded.[13]

Early Heritage Recognition

The iron and steel company Avesta Jernverks AB not only built and used the industrial area along the river for the company's production, it was also the first to interpret it from a historical perspective. Industrial heritage researcher Marie Nisser has shown how certain industrial sectors in Sweden, most notably the mining and iron and steel industries, worked consciously during the whole of the twentieth century to establish archives and preserve their older built environment.[14] The owners of Avesta Jernverks AB—successive heads of the Johnson family—were the kind of industrialists who invested a lot of effort in establishing archives and preserving several of the various premises within their group of companies.

Avesta Jernverks AB and the Johnson family also commissioned one of their employees, the engineer Bo Hermelin, to be in charge of "cultural matters" at the company. Hermelin collected old, no longer used artifacts from different industrial sites owned by the company, and his ambition in the 1940s was to establish a mining and metal museum in Avesta. The idea was to show the development of the company and iron making in general "from oldest times to the present day" from an international perspective.[15] In parallel to the town center proposal the architect Alvar Aalto was asked to design a research institute for the Johnson group of companies, including laboratories, offices, and the museum. The institute was to be built just outside Avesta and the museum was to occupy a substantial, centrally located space in the institute complex.[16] However, similar to the town center proposal, the research institute and the museum were never realized.[17]

Parallel to the company's heritage efforts, professional heritage recognition of industrial production developed on the national level. During the first half of the twentieth century a number of museums focused on the history of technology and industry were inaugurated.[18] This early interest was principally directed toward one single sector, the iron industry, and originated in the history of certain companies.[19] Thus, at this time both company representatives and heritage professionals exhibited an interest in industrial history and heritage, even though not all the plans in Avesta were fulfilled.

In the 1960s a more general interest in industrial history emerged, based on a perspective "from below" combined with an interest in industrial architecture. Appeals were formulated that aspired to bring about "volunteer inventory work of buildings and milieus connected to work," inspired by such activities already going on in Britain.[20] Among historians, architects, and building conservationists, an interest in older industrial buildings had slowly grown, and the topic was discussed at professional conferences and in the press.[21] By that time industry had begun to be understood as something about to disappear, something that had to be saved before it was too late.

In the 1970s, the management of Avesta Jernverks AB was still planning to establish a museum, but instead of a new building the intention was to convert the old blast furnace plant, which had been abandoned three decades previously, to house the museum.[22] Bo Hermelin wrote to the executive director about how he imagined the realization of the museum: "In general, all that conceals the architecture, the powerful walls and the beautiful arches, will be taken away, unless the function is obvious or interesting in other respects, or perhaps has a picturesque appearance."[23] Hermelin further suggested that the selection and display of suitable objects should be undertaken in consultation with the director of the National Museum of Science and Technology in Stockholm. Avesta Jernverks AB thus worked actively and ambitiously to create a museum. This also meant that the old blast furnace plant would be preserved, although it did not fill any function in the company's production. On the initiative of the company and some local politicians, the old blast furnace plant was declared to be of national interest to heritage conservation in 1987.[24] However, not even at this moment did the envisioned mining and metal museum become realized in practice.[25]

The European Bison

Avesta Jernverks AB and the Johnson family certainly found historic consciousness to be an important issue for the company. Another matter that engaged the industrialist family was the bison.[26] During the first decades of the twentieth century, several zoos began to house European bison since the species was endangered in the wild. The Swedish zoo and open-air museum of Skansen in Stockholm was one of these, but the available space at Skansen soon proved too small. In the 1930s, Axel Ax:son Johnson at Avesta Jernverks AB suggested that a number of bison be moved to Avesta and cared for by the company. Skansen agreed, and about ten live animals were moved to a fenced park at the outskirts of the town. The company employed a special bison keeper, a position that was later inherited from father to son throughout the twentieth century, indicating its willingness to accept long-term responsibility for the bison, analogous to its responsibility toward the local community.

In 1952, the bison was incorporated as a fundamental element of the company's logotype, and soon it became a valuable trademark. The bison was said to represent key characteristics of the stainless steel produced in Avesta—strong and imperishable—and for more than three decades the company employed an artist to create sculptures and fancy goods in stainless steel, with the bison as the most frequent theme. Avesta Jernverks AB also brought customers to the bison park to see the impressive animals. Such visits, together with a dinner of roast bison and perhaps a gift in the form of a small bison sculpture, were supposed to sway customers in their choice of an iron and steel contractor.

In 1976, when Avesta Jernverks AB was to inaugurate a new sheet rolling mill at the Southern Works, a large stainless steel sculpture was unveiled and placed on a small hill just outside the industrial area, clearly visible from the road that carried most of the traffic passing through Avesta. The sculpture—a male, muscular bison, slightly larger than its living model—was named "Stan," alluding to the English term "stainless," and the company artist Lars Andersson had worked for several years to complete it. Until 1989, when a new through road was built, the bison sculpture "Stan" was a landmark for the town. Later in this chapter we will return to the story of the European bison in Avesta, since it in many ways epitomizes the changing relationship between the company and the town, and in its sequel, the enduring spirit of the company town.

The Spirit of the Company Town

The company town of Avesta—a Swedish "bruk"—grew and established gradually over several centuries. Simultaneously, the power hierarchies and mutual responsibilities defining the roles played by company management and workers became firmly established, and the spirit of a company town permeated the way of life. In general, on the positive side, the spirit of the company town was associated with loyalty, professional pride, fellowship, and safety. On the negative side, it was linked to social injustice, passivity, conservatism, and jealousy.[27] The physical structure of the town illustrated the different relationships, and in many ways the company owner acted as a father figure in the community—unquestionably superior, but with the ambition to do what he regarded as best for the town and its residents.[28]

The Johnsons were, among other things, intimately involved in town planning issues, especially housing, in local politics, in the church, and in the lives of individuals. The Avesta Jernverks AB not only donated substantial pieces of land to the town on the condition that the town build dwellings for workers and paid for an indoor swimming pool and several town festivals, it also headed a vocational school. As we have seen, it also took care of history and heritage matters and kept a small bison farm both to protect the species and to uphold symbolic values. Other industrial enterprises met with substantial resistance if they tried to establish a foothold in Avesta, since the Johnson family saw them as competitors, potentially threatening the dominance of Avesta Jernverks AB.[29]

Economic historian Therese Nordlund shows that it was a conscious strategy of the company management to "make the town dependent."[30] One of the key members of the family, Axel Ax:son Johnson, even wrote in a letter that "Avesta is my private property and I treat it as such and wish to follow the course of events taking place there."[31] The dependency was nevertheless seen as mutual, with local politicians arguing that "Avesta turns on the success of the ironworks" and that the company and the town "live, prosper and die together."[32] Hence, in the company town of Avesta, the company's entanglement with the town extended far beyond issues directly related to the iron and steel production. What was the future of this social contract?

A New Owner of the Old Industrial Area

During the 1970s the iron and steel sector faced severe challenges. However, the crisis initially hit the production of ordinary steel, so that in Avesta, where the main production was special steel, no major realignments and downsizings were necessary until the early 1980s. While Swedish producers of ordinary steel were unified in Svenskt Stål AB (SSAB), the companies that produced special steel, including Avesta Jernverks AB, underwent a similar, step-by-step unification, including fusions with foreign companies.

In the mid-1980s, the company decided to concentrate all units in the Southern Works, sold the Northern Works to the municipality of Avesta, and then stepped back and left its original location without any substantial ideas for its future use.[33] The contract between the company and the municipality simply stipulated that any underground contamination to be dealt with according to environmental legislation was the responsibility of the company, while the municipality was to undertake the task of clearing the pollution above-ground.[34] At this time, the company had reduced its workforce from a peak of thirty-six hundred employees in the late 1960s to about one thousand, and at the same time the town's population decreased slightly, to about eighteen thousand.[35]

In spite of the expected constancy in the relationship between the industrialist family and the town of Avesta—from the perspective of the Avesta inhabitants, after one of the several mergers the Johnson family sold its shares and vacated Avesta as one of their places of residence in 1992. All at once, the interest of both the family and the company in the local community virtually evaporated.[36] An epoch came to an end, although the family did keep the manor, located close to the Northern Works industrial area, for private use.

The municipality of Avesta became the new owner of the Northern Works, with the intention of putting the partly derelict industrial area in order and trying to attract another major industrial company. Upon the municipal takeover of the Northern Works, every possibility to increase or stabilize the number of workplaces, especially for men, was regarded as significant for the future prospects of the company town.[37] The industrial area, however, soon showed a mix of empty buildings and premises leased to smaller engineering industries and associations. Among others these included the hot rod club, Steel Town Cruisers Avesta, and the Avesta Kennel Club, along with companies working with energy insulation and metal cutting, while some locations housed publicly financed activities, such as the recycling of old furniture and household goods, a youth club, and government-subsidized workplaces. Furthermore, private persons and companies rented storage space in the former industrial buildings.[38] Some of the buildings were comparatively well kept while others were falling into decay. Many windows were broken or boarded up, and buildings that had no tenants offered a tempting refuge for youngsters. The long road through the area was used for illicit street races.

A municipal real estate company, "Avesta Industristad," was founded for the express purpose of managing the Northern Works, and one of its first initiatives was a competition in which the inhabitants of Avesta were invited to propose a new name for the area. From the almost six hundred proposals received, the jury chose the name "Koppardalen" ("the Copper Valley").[39] The new name alluded to the copper works, the first industrial enterprise at the site, and the coin manufacturing that constituted part of a proud history, which, besides the prestigious task of national coin production, included spectacular orders of copper plate for the roof of Versailles Palace outside Paris.

Nonetheless, efforts to find a new big company willing to establish production in the redundant industrial area were unsuccessful, and the municipality eventually had to redirect its goals. The previous state of affairs, with the Johnson family clearly directing important development steps, no longer existed, and proved impossible to replace with a similar industrial actor. Instead, the public body of the municipality had to step forward and explore other potential future imaginaries without the company management as the driving force and obvious leader of the town. The Johnson father figure had cut the close connection and left it to the residents of the town to redefine their role in relation to the former industrial area. The patriarchal structure was disrupted and the social contract broken, rending a wound in the community. The abandoned industrial area in the midst of the town had become a physical reminder of pain. Who should be in charge of delineating the new meaning and the possible new use of this piece of land in order to bring a new future direction to the local community?

A "Cultural Team of Workmen" and a Municipal Project

In the 1990s the search for new ways to use the Koppardalen area was launched with pragmatic investigations of the physical condition of the old blast furnace plant. It was soon verified that the roof was leaking in several places and in need of considerable repair. Furthermore, the building had many broken windows and the floors were covered with thick layers of pigeon droppings. Avesta Industristad received employment subsidies from the government for an initial restoration, but even after the roofs and windows had been repaired and the interior somewhat cleaned, the impression of a dirty, ruined place remained. "When you were there, you could not stay for long because you got totally black from dust" was a common comment.[40]

However, in 1993, it had been cleaned enough to house a theatrical performance, which took advantage of the setting in the plant by using real fire in the furnaces, light, smoke, and sound effects. The producer stated that the industrial environment was a valuable asset due to its "fantastic, magic atmosphere."[41] Two years later a handful of people from the cultural sector realized a long-cherished dream about an exhibition of contemporary art in the plant.[42] The exhibition was named "Avesta Art" and inaugurated with a program called "The Blazing Hearth of Our Ancestors." In the inauguration program, an actor dressed as an old ironworker described the work in the blast furnace plant; one of the initiators, writer and politician Karin Perers, suggested that a new team of workmen had thus taken charge in the old blast furnace plant—a "team of cultural workmen."[43] The exhibition was a huge success and became a recurring event that continues to attract large audiences (figure 6.2). The built environment was regarded as one explanation for the triumph of the exhibition, since "contemporary art in interplay with shimmering slag stone and powerful furnaces give birth to unexpected encounters as well as magic adventures."[44] The very blast furnace plant was called a "cathedral of work," an "Inca temple" and a "medieval castle."[45]

In retrospect, one of the other initiators, the head of the department for cultural and educational matters, Lars Åke Everbrand, recalled the feeling when planning the art exhibition: "It was as if you

Figure 6.2 The old blast furnace plant was built in the 1870s out of slag stone brick, a by-product from the ore melting process. In the 1990s, the plant opened as a gallery for contemporary art. Photo: Jan af Geijerstam, 2003.

describe yourself as a great person...ashamed to say so...how are we to do anything international with contemporary art...down there in that rubbish...what will people think of this?"[46] His description illustrates the presence of the spirit of a company town, according to which it is always wrong to praise oneself. Nevertheless, Everbrand was sure the exhibition was going to work, in spite of the assumed negative judgments from the local community.[47]

These two cultural events, the theater play and the art exhibition, were the first steps on a long and winding road toward a new significance of the former Northern Works, now the Koppardalen area. The municipality took the lead in the transformation by forming a project team assigned the task of working on the development of Koppardalen. At the beginning of 1997, the project team proposed dividing the industrial area into three sections, based on existing characteristics and as an expression of future plans. The western section, with the old blast furnace plant, was to become a place for mainly cultural and educational activities, while the central section, with a mixture of buildings from different periods, was planned to accommodate smaller companies within the service sector. In the eastern section, with modern industrial plants, continuous industrial production was to be retained, at least for the time being. The project team presented a proposal that implied a five-year development plan divided into two phases. The overall idea was to integrate Koppardalen with the existing town center and to create a more attractive center that "could give a strong local identity to Avesta and its population, to contribute to making Avesta known within a larger region, and to attract both visitors and entrepreneurs."[48] Financially, the municipality was able to attract economic subsidies from the government as well as European Union funding directed toward industrial regions in decline to realize its plans for the Koppardalen area.[49]

In connection to the plans, municipal representatives bemoaned Avesta's failure to realize the architectural proposals by Alvar Aalto: "If these...projects had become reality, they would not only have unanimously changed the townscape but would also have made the town richer in tourist attractions. They would probably, in a decisive way, have influenced the life of the inhabitants in Avesta."[50] A strong image of the town as it could have been was lamented.

Nevertheless, in Koppardalen there were existing architectural and historical treasures to explore and expose. The old blast furnace plant, designated as of national interest from a heritage perspective in the late 1980s, was perhaps the most obvious object in this respect. Other buildings in Koppardalen built in the glassy blue, green, or black slag stone brick were also recognized, as were some buildings designed by renowned architects like Ivar Tengbom and Torben Grut. These buildings were, however, mostly encapsulated by newer buildings, creating a densely built environment with narrow corridors for people and vehicles. Moreover, because of the previous direct use of water power for industrial production, the riverbank was almost completely inaccessible from land. Thus, the identified treasures of Koppardalen were understood to be hidden, waiting to be rediscovered.

Consequently, the Avesta municipality decided to carry out some demolition for the purpose of increasing accessibility to the buildings and locations that were regarded as most valuable.[51] First on the list was a sheet rolling mill, built in clay brick mainly in the 1940s, but with additions from the 1960s, described by the County Antiquarian Ulf Löfwall as of a "universal design that is not significant for the distinctive character of the area."[52] Generally, the municipal project team stated that converting some of the buildings was just as important as getting rid of "ugly, dilapidated milieus [to thus] increase access to more valuable buildings and to enhance the contact with the river."[53] The very process of demolishing the sheet rolling mill was slowed, however, by the decision to entrust unemployed workers with the job, who were paid by governmental subsidies on an irregular basis. Over the six years between 1994 and 2000 the mill disappeared, one piece at a time.

In the new local valuation of the Koppardalen area, aesthetic features became predominant, along with conceptions of authenticity. As to the built environment, slag stone brick and a location along the river were regarded to be most attractive. A few opposing voices were raised from actors outside Avesta, but the local view was never seriously contested.[54] However, concerning issues of cleaning and greenery, contradicting ideals were expressed by local citizens as opposed to the municipal authorities. According to public opinion, cleaning was regarded as necessary so that the buildings could be used for new purposes. From an antiquarian perspective, however, it was argued that "all cleaning should be related to the authentic working environment" and thus avoided, since it damaged a genuine experience.[55] Consequently, in hindsight, Lars Åke Everbrand regrets the decision to put a molded floor in such a large part of the old blast furnace plant. Today he thinks they should have preserved a corner as a point of reference to show how it looked in the late 1980s.[56] In a similar manner, many envisioned greenery to enhance the experience of the area. One Avesta resident, for example, suggested to "plan the area so it becomes nice [with] green areas" and "take away all the ugly, grey and brown houses [and instead make] oases and other walking areas." Other residents suggested that one should "tear down the slag stone buildings and build dwellings with a view of the river, take away the asphalt and establish green areas."[57] On the contrary, from the antiquarian perspective, trees and bushes that had spontaneously started to grow in Koppardalen during the decades of abandonment were said to "give the area a woody quality that is unfamiliar to its character."[58]

The resulting strategy must be described as a compromise. It was decided that Koppardalen was to be "developed through planting," while vegetation in the direct context of industrial buildings and structures was to be removed.[59] In spite of the different valuations expressed regarding the visual appearance, the local vision was largely cohesive. Koppardalen was imagined as a lively and appreciated place where, as in the old days, thousands of people had their workplace and where, for the first time, the present town center could find itself extended into the waterfront lowland that had previously been closed to public access.

The first phase of the municipal development project in Koppardalen was completed in 2000 at a cost of about forty million Swedish crowns, of which the municipality had contributed twenty-five million. This meant it was the town's largest single project in terms of money. During these years it had torn down the previously mentioned sheet rolling mill and parts of a cold rolling mill, decontaminated land of mercury, repaired and put new roofs on a number of buildings, built a pedestrian and bicycle bridge between Koppardalen and the town center, and, finally, almost finished construction of a sports arena. Toward the end of the first phase, a clear majority of the Avesta inhabitants were positive about what was termed "the renewal" of Koppardalen.[60]

There were also voices raised against the changes, sometimes in anger, suggesting that the buildings in the area should be forgotten, torn down, or burnt. One person wrote in the local newspaper *Avesta Tidning*:

> The municipality's buying the Northern Works industrial area and the conglomeration of unprofitable littered buildings at a cost of 36 million kronor plus interest plus undertaking of demolition and clearance plus maintenance of the heritage listed buildings for ever and ever... could something worse fall upon a small town like Avesta? This is crazy... The inhabitants of Avesta have had enough of the old ironworks, you do not go there in order to feel at home. It is a narrow dirty blind alley in history.[61]

Still, we can conclude that, on the whole, municipal leaders found public support for their activities, and that public objections to the transformation process in Koppardalen focused mainly on the municipal economy.[62] Because the project was largely financed by taxes, the activities had to compete politically in the same arena as, for example, the school and the hospital. The symbolic value and the kind of new possible uses of the area were thus not a major issue for the public at large. However, to a smaller number of individuals, the transformation of the Koppardalen area became their main professional mission.

The New "Acropolis"

The Northern Works industrial area had been abandoned, both in a physical sense and in terms of identification by the main actor: the industrialist family, the Johnsons. After a period in limbo, the municipal public body, a bit hesitantly, took over responsibility for the area and gradually turned it into a key focus for the future of the town of Avesta. But who actually were the individuals and what were their motives?

The people who were most deeply engaged in the transformation of Koppardalen were those who had first experienced the area in its abandoned and derelict state, that is, they had not worked there when it was in production. They were primarily municipal civil servants or employees at the municipal real estate company, the county museum, or the county administration. The second phase of the municipal project concerning Koppardalen began with two people, Lars Åke Everbrand, and the town architect, Dan Ola Norberg, who were commissioned to formulate ideas for the continuation of the project.[63] Their proposal was presented in the spring of 2001 and described two main alternatives, one highly ambitious and the other less so. The less ambitious alternative was presented first, proposing a nightmare scenario with broken windows, a decreasing population, and increasing unemployment.

The more ambitious alternative, on the other hand, depicted a future that was different in all essential aspects. It envisioned international conferences taking place in Koppardalen, where big service companies were already established and students enthusiastically gathered around exciting technical models in a visitors' center. The greater ambition was connected with the ill-fated Acropolis, Alvar Aalto's plan for a new town center in the 1940s, and the future Koppardalen was called "the new Acropolis." The authors even warned against repeating the mistake made in the 1940s when the town center design came to nothing. Everbrand and Norberg asserted that the future use of Koppardalen concerned nothing less than the future of the town of Avesta, asserting polemically: "The question is not whether we can afford to invest, but whether we can afford *not* to invest."[64]

The municipal council also decided to increase the budget in order to finish the activities begun during the first phase, and to continue the project into the second phase.[65] The decision to launch the second phase did not, however, imply unconditional acceptance of the analysis formulated by Everbrand and Norberg; elements were both added and removed. The biggest investment plans during the second phase became the conversion of a plate rolling mill and a sheet rolling mill into mainly office space, and the development of "historic milieus" in the old blast furnace plant, respectively. The plans also included pilot studies for an indoor swimming pool, an educational center, and a science center.

Experience shows that the existence of one or a few truly dedicated persons has been crucial to the feasibility of many similar transformation projects, especially during the formative stages.[66] Lars Åke Everbrand's ambitions were high from the outset, or in his own words, he personally carried "a rather great dose of aspirations," describing himself as a competitive kind of person, always aiming to create something innovative.[67] In retrospect he does not regret anything and if he had to repeat it, he would not change his decisions. "I wanted to do something that would be noticed and worth something," he said, comparing the work of renewal in Koppardalen with the daily work of municipal administration.[68] He emphasized how he and a few other people had their hearts in Koppardalen and that this made them very insensitive to working overtime. Everbrand described the possibility of realizing new and innovative ideas in Avesta in terms of freedom and, economically, as "a kind of economic dictatorship." However, he expressed concern about the project possibly becoming more institutionalized, and about personal devotion colliding with an administrative form. Then he believed it would feel "meaningless."[69]

Step by step, the industrial land of the Northern Works turned into the post-industrial landscape of Koppardalen, characterized by visual consumption and explicit historic references. A future vision was gradually taking shape and shared by a majority in the local community. The degree to which the consensus succeeded, or the spirit of the company town endured, was illustrated by the conflict that arose in connection with the traces of the aforementioned sheet rolling mill.

In the visualized renewal of Koppardalen, the sheet rolling mill had been located on a conceptual borderline within the industrial area.

The mill was physically linked to the building complex of the western section established by the municipal project team, but belonged to the central section in the transformation plans. Its existence could thus be seen as a question of negotiating between the endeavor to create cultural and educational activities and the efforts to provide commercial premises for smaller companies.

During the demolition, Dan Ola Norberg, Lars Åke Everbrand, and the County Antiquarian Ulf Löfwall discussed different ways of allowing the sheet rolling mill to leave some traces on the site. They argued that to "understand the historic cultural heritage from the copper works period up to today it is as important to protect the traces as to preserve certain buildings. It is by being able to read the history in the material environment that we can tell and explain the daily life at the works and the development of industry."[70] Their argumentation ended in a decision to leave a row of iron girders from one of the long sides of the sheet rolling mill. Another trace they prioritized was to leave an exposed gable—a former interior wall—untouched (figure 6.3). The gable was seen as a piece of art with a multiplicity

Figure 6.3 As one part of the transformation of the old industrial area in Avesta, a sheet rolling mill was torn down. It was decided to keep some traces of the mill, among them a row of iron girders from one of the walls (to the left in the picture). The decision caused indignant reactions, however, so that the iron girders were later cut down. Photo: Kent Lindström, 2002.

of colors and patterns originating from earlier floor levels, a staircase, and partition in rooms. As an addition to the traces preserved, they also decided to create a new material reference to the removed sheet rolling mill. This was done by framing a number of new parking lots with low net cages filled with bricks from the demolition.

However, not everybody found the iron girders attractive; some said they were reminiscent of a concentration camp. The managing director of Avesta Industristad, Jan Thamsten, was one of those actors who regarded the remaining traces of the sheet rolling mill as strange and ugly features in the environment. In his eyes, the row of iron girders and the colored gable became obstacles during meetings with potential tenants for the area, something that required explanation and did not benefit the marketing of the location. When the oil company OKQ8, after several years of discussion, decided to move its customer support for the Nordic countries from the Swedish capital, Stockholm, to Avesta, and into a new office space in an old plate rolling mill in Koppardalen, it was regarded as a great success from the municipality's point of view. The oil company brought about forty new jobs and—probably most importantly—legitimized the public money spent on converting the buildings to new use. From OKQ8's point of view, the prospect of finding a loyal workforce for what was commonly considered a transition job was central. In connection with OKQ8 moving into its new premises in spring 2003, Jan Thamsten decided that the iron girders had to go. After conferring with the County Antiquarian he promptly executed the decision without consulting others in the municipal leadership. The iron girders were cut down and the gable was plastered into a single, homogeneous, gray surface.

The same weekend, the chairman of the municipal council was about to show some council members and a couple of guests the ongoing transformation of Koppardalen. The chairman, the Center Party member Karin Perers, summoned the group to the row of high vertical iron girders in the middle of the industrial area. When they arrived at the designated place there were no iron girders to be seen. Astonished and dismayed, Perers realized that the iron girders had been hastily removed without her knowledge.

At the time when the iron girders suddenly disappeared, Perers had been chairperson for only a couple of months. The visit to the old industrial area of Koppardalen was a first step in her ambition to bring new knowledge and ideas to the members of the council. The guided walk through the industrial area was designated as a "study visit," and followed by an extended council meeting a few days later.

Perers had invited special guest speakers who were to give their views on the future of Koppardalen, together with representatives from all the political parties in the council. At the meeting, none of the speakers made an explicit issue of the removed iron girders; in spite of the fact that most of them had taken part in the weekend study visit.[71] The event was presented in the local newspaper and the municipal council meeting was broadcasted live on the local radio channel. The main impression is that all actors were eager to smooth over the incident of the removed iron girders and the plastered gable. The overriding goal of renewal of the old industrial area could not afford too many explicit internal conflicts. The incident was not regarded to be absolutely crucial, although several of the actors still commented on it with frustration years later.[72]

Thus, the cohesive vision of the Koppardalen area was sustained within the municipal organization, a fact that was further emphasized by an exceptional level of political agreement as regarded the transformation process. During the remarkably long period between 1919 and 2002, the Social Democratic Party held a political majority in Avesta. After the election in 2002, a coalition consisting of the Conservatives, the Center Party, the Green Party, the Liberal Party, the Christian Democrats, and the local party Axel Ingmars Lista—Avestapartiet, came into power. However, the direction of the municipal project in Koppardalen was not changed.[73]

Shaping Local Views

Were there any other voices in the transformation process that challenged the municipal views? What about the company, what about the former workers, what about other ideas for directing future changes in Avesta?

There were, for example, two groups of retired iron and steel workers who met regularly every week from the mid-1980s on to talk about memories of their working life and about contemporary society. However, they did not harbor any ambition to influence the course of events at their former workplace or to participate in defining how its history was to be written. This might be a general picture in Sweden where, according to Kjersti Bosdotter at the metal union in Sweden, former workers in mining and metal industries have seldom shown much interest in telling their stories.[74] This can perhaps be understood as alienation from or a difficulty to identify with the transformation process and the aestheticization of former industrial milieus from a heritage point of view.[75]

Other potential actors in the transformation process were a local guide association and a local history society formed in Avesta in the late 1990s and early 2000. Apart from being established relatively late, these organizations were not radical in the sense that they insisted on the primacy of a workers' history "from below." While economic historian Maths Isacson pointed to an increasing general interest in local history at the turn of the twenty-first century due to the vanishing industrial society, and also indicated that voluntary activities by local history associations tended to be associated with the reuse of industrial buildings as museums and art galleries, this link cannot be confirmed by the case of Avesta.[76] Instead, the voices "from below" are remarkably absent in the transformation of the Koppardalen area, although the driving forces in the municipal administration did indeed take on the social ambitions of the industrialist family of the Johnsons—to do what was regarded as best for the local community and its people.

To strengthen this picture even more, the municipal project team and local politicians explicitly stated their superior position in terms of knowledge, implying their ability to make good decisions for the benefit of the supposedly more ignorant citizens of Avesta. In connection with the transformation of the Koppardalen area, for example, the local councilor commented on difficult decisions by saying that "you have to be brave and push things when you have knowledge and perhaps another kind of input, which influence the decisions you make, compared to those who just read the local newspaper."[77] "Those who just read the local newspaper" were nevertheless also the objects of educational efforts in relation to the new activities in Koppardalen. For example: What should be regarded as historically correct in terms of architecture; what were the characteristics of the international arena of renewal projects in which Avesta wanted to play; and what was the appropriate way of viewing contemporary art? From the municipal leaders' point of view, the population of Avesta had to learn more about these issues.[78]

To aim for change that was to become a public good was another way of expressing this ambition to do what was best for the town inhabitants. The architect Jan Burell, for example, emphasized that the effort to make Koppardalen a place of public access was very important to him, such as building an indoor swimming pool in one of the former ironworks buildings. When plans to build the indoor pool were scrapped, Burell expressed his concerns about other future plans for the industrial building. One idea was to convert it into housing, an idea he found "exciting, but then in the worst case it could become private property," something he apparently wished to avoid.[79]

Avesta and the Koppardalen area were compared to other places in the country and abroad in order to get ideas and inspiration, but also to convince a hesitant local public at home. Among many travels, a visit to transformed industrial areas in England was, according to Lars Åke Everbrand, "incredibly important, because then they saw, then they accepted. They had not visited any place like that before and we were able to see a development that they could compare to Koppardalen, how Koppardalen could develop in the same way. That was important: We are part of an international trend."[80]

It is clear that informal groups were able to work unobstructed, initiate ideas, and carry them through. However, external recognition was important, for example, when one of Sweden's biggest television news programs, *Aktuellt*, reported about the art exhibition in the old blast furnace plant during peak viewing hours. Everbrand asserts that as soon as they received this nationwide exposure, local politicians also regarded the exhibition as terrific.

The ease with which the transformation work in Koppardalen proceeded could largely be explained by the reliable communication that had become established within the group of leading actors. When something had to be decided or executed in a hurry, a simple phone call could solve the problem. This group, mostly comprised of men, included not only longtime acquaintances, but people who had also long held the same kinds of professional positions in their organizations—in some respects, quite similar to an industrialist family.

What are the changes in Avesta and Koppardalen an example of? One thing that is clear, I argue, is that the transformation of the Northern Works industrial area challenges sometimes simplified presumptions about how former industrial areas have generally turned into post-industrial landscapes of cultural and commercial arenas, especially in two respects: One is the presumed connection between the 1960s and 1970s interest in history "from below" and the more large-scale physical transformation projects—a connection that is refuted by the course of events in Avesta. In Avesta history, change and continuity have always been mainly generated "from above"—by an elite. Another observation is the differences between transformation projects carried out in the countryside or comparatively small towns—like Avesta—in relation to the reuse of industrial areas and buildings in larger cities. In Avesta the changes were rather slow, evolved gradually, and were headed by a local public body, publicly financed and connected with the very survival of the town. In many other investigated transformations, the changes were instead fast, headed by private actors, driven by high development pressure and likely economic returns. In the slower publicly driven transformation the spirit of the company town endured, while the other type of process generally lacked long-term engagement with the local community.

The European Bison, Part Two

Now let us return to the bison in Avesta. What happened to the living animals in the fenced park outside the town and to the large bison sculpture "Stan," which was unveiled in 1976 upon the inauguration of a new sheet rolling mill at the Southern Works and soon became a landmark in Avesta, mainly associated with the Johnson company and patronage? In 2001, when the company merged with Finnish Outokumpu Steel, the bison disappeared from the logotype. The company also removed the sculpture from its hill, which caused an indignant reaction from Avesta residents. The local manager assured that the bison sculpture was in safe keeping, however, and would to be placed elsewhere in the town.[81] In consultation with representatives for the town of Avesta, the company—now called Avesta Polarit—decided to move the sculpture to a prominent place firmly associated with the municipal administration of the town—the square between the town hall and the shopping street (figure 6.4). It was distanced from the company realm, but the company stressed that it still owned and cared about its "baby."

On night in December 2001, "Stan" was placed on a newly built pedestal of dark green slag stone and wrapped in paper and gift ribbons. Again, the sculpture was unveiled with festivities, paid and organized by the Avesta Polarit company, which presented the bison sculpture as a Christmas gift to the inhabitants of Avesta. On a signpost beside the sculpture it was stated: "To the inhabitants of Avesta the bison became a symbol for continuity and confidence in the development of Avesta community... In 2001 the bison disappeared from the company logotype, but in the hearts of the people of Avesta it will always remain." The name "Stan," alluding to "stainless," could a few decades later easily be understood as an allusion to "stan," which in Swedish is an abbreviation for "staden," meaning "the town." Hence, just as the Northern Works industrial area had been transformed into the post-industrial Koppardalen owned by the municipality, the company's trademark symbol of the bison had been transferred to become a symbol for

Figure 6.4 The company symbol, a bison sculpture in stainless steel, initially had its place just outside the ironworks. In 2001, the new owners of the company decided to donate the sculpture to the town. The bison was moved to the prominent square between the town hall and the shopping street. Photo: Anna Storm, 2003.

the town. "Stan" was soon also incorporated as a prominent part of a set of trademark images used by the Avesta municipality on its website, on maps, and tourist brochures, along with motifs such as the old blast furnace plant, the Dalecarlian horse, and the Avesta speedway star Tony Rickardsson. Furthermore, the local councilor believed it would be a good idea to begin producing small bison sculptures for sale and as official gifts. The bison representative at the company agreed and thought the bison would be perfect as a "municipal animal" in Avesta.[82]

The living bison in the park were also repurposed, in part as a way to attract new inhabitants and new companies to Avesta. For example, the 2002 story about how the female bison calf "Avebeta" was thriving was used to symbolize how the town of Avesta was growing strong and robust.[83] Formally, the company still owned the bison but was not interested in "managing a zoo," as the Finnish CEO Pekka Erkkilä said. Instead, it began negotiating with the municipality about a change in ownership.[84] The positive symbolic associations with the bison, along with economic and practical responsibilities for a sculpture and a park, were thus taken over by the town and the municipality of Avesta. It is apparent that a symbol can remain constant even as its meaning transforms.

One symbolic institution that could have been realized, but was not, was the mining and metal museum initiated by Avesta Jernverks AB. At the beginning of the twenty-first century, the items collected by Bo Hermelin half a century before were still left under a thick layer of dust in a decaying industrial building in a remote corner of the Koppardalen area, with classification tags intact. Were they awaiting discovery, or did they perhaps not fit into the new conceptualization of the area and the future of the town?

Concluding Remarks

The enduring spirit of the company town has been acknowledged by several researchers, especially from a perspective of how to make change possible in supposedly inert and slowly fading company towns. Human geographer Gunnel Forsberg, for example, asked whether there was an "equal" version of the spirit, transcending the power hierarchies linked with work and gender.[85] The company towns are in many respects "male" communities in their organization and valuation of capacities, and thus constitute a remnant of an obsolescent societal convention.[86] Would it be possible to keep and emphasize positive aspects of the spirit of a company town, such as fellowship and safety, and at the same time leave out the negative parts? Or would it be best do everything possible to fight this persistent spirit?

From most perspectives the transformations that took place in Avesta were viewed as a success. From a national point of view, it became an example of the progressive renewal of a former industrial area, adhering to contemporary buzzwords like small-scale entrepreneurship and cultural activities as keys to strengthen local identity and enhancing a tourist-based economy. From a local outlook, the decades marked by uncertainty in leadership gradually found a new focus in the joint efforts to change the Koppardalen area. The town and the municipality made it the biggest investment for the future, both in terms of monetary resources and in symbolic currencies. The industrialist family was replaced by a fairly informal but very tight group of mostly men, who shared a vision of a transformed industrial area to benefit the town and its inhabitants. I have argued that this transformation process signifies one way in which the spirit of the company town has endured; had it been a conscious strategy, it could have been articulated as an ambition to "stay the same by means of change." The post-industrial situation thus comprised both new and familiar features to the Avesta residents: the content was new, but the form, in many aspects, remained. In short, a "new team of workmen" had begun their shift.

So should this be considered something good? Is the enduring spirit of the company town of Avesta a sign of a properly healed wound, or rather an unfortunate clinging to outdated societal structures that represent artificial respiration at best? Is the predominant consensus culture a mark of strength in a small town, or rather of the adoption of suppressive power hierarchies? Probably not surprising, I believe it is both. As noted in the introduction, a scar remains ambiguous, not least so when the main aspect of this particular post-industrial landscape scar is something as amorphous as the spirit of the company town. While to some people the Koppardalen area had turned into a post-industrial landscape of visual desire and emotion, it remained industrial land for others, although projections of the kinds of work carried out there were modified to include office jobs. Finally, to some people, the long narrow piece of land along the river in Avesta had become a kind of nowhere, representing a part of history not intertwined with their contemporary life. The shimmering dark green slag stone walls in the old blast furnace plant were certainly beautiful, but the former iron workers did not go there anymore.

Chapter 7
Prospective Scars Unfolding

The five places and stories we have encountered in this book each deploy one critical aspect of the post-industrial landscape scar. The scars have come into view as unstable and sometimes in motion, they are defined by different geographical scales and perceived distances, they tell of losses and betrayed dreams, they highlight ideas about nature and organic growth as a key future imaginary, and they make visible persisting spirits of community as well as repressive hierarchical structures.

The experience of decisive and painful changes—which turned these places into post-industrial landscape scars—is paralleled by transformed yet continuing social contracts and ways of doing things. The enduring spirit of the company town in Avesta (chapter 6), expressed in the replacement of the industrialist family by municipal leaders, echoes in the "spirit of Barsebäck" (chapter 3) carried by the nuclear power plant employees into the period of working in a closed-down industry, denoting both friendly comradeship and joined forces toward the external enemy, namely, the distrusting antinuclear publics and politics. The unquestionable dominance of the mining company LKAB in the search for future prospects in Malmberget (chapter 2) remains, as does the dependency on supranational leaders in Sniečkus/Visaginas at the Ignalina nuclear power plant (chapter 4), leaders positioned far away in Moscow or Brussels. The long established power hierarchies in Malmberget and Sniečkus/Visaginas are mirrored by chastened attitudes in the local communities and make up part of the explanation as to why there is a lack of agency and articulated shared opinions in these places.

The capacity to act, or the lack thereof, also relies on time elapsing and on tradition. An experience of betrayal is shared by people in all five places, but a turning point has clearly been passed in Duisburg (chapter 5) and in Avesta—which is one of the reasons why I categorize these places as "reused" post-industrial landscape scars. Malmberget, Barsebäck and Ignalina, are still teetering at the brink of renewal, which is why I describe them as "undefined" scars. The two nuclear power plants of Barsebäck and Ignalina no longer produce electricity, but even now they constitute active workplaces, preparing for waste management and the dismantling of the plants. In Malmberget mining is still under way, albeit with highly uncertain future prospects. The industry in these places does not yet clearly represent the past. The situation is liminal, like a scab, vulnerable and easily turned into a wound once more. Scarring and healing takes time, which is not to say that Malmberget, Barsebäck, and Ignalina will tread the same path as Duisburg or Avesta, but only that the latter have spent more time on recovery and a search for alternative futures.

The fact that Duisburg and Avesta are based on a history of the iron and steel industry makes them special in a sense. The leading players in the iron and steel sectors have long valued engagement with the past and acknowledged a social responsibility for their dependent communities. This tradition has probably affected the local response to the dramatic changes of the late twentieth century. When industrialists and owners formally withdrew from their leading positions in the community context, their role was taken over by public actors, so that the social contract in some respect remained. In contrast, because the young nuclear power sector and quickly established mining towns like Malmberget lack this tradition, they are less stable, or less inert. In Malmberget, the prevailing economic rationale made it easy or even natural to let the Pit expand closer and closer to people's homes and to abandon or demolish key public buildings.

Instability and insecurity are most apparent in Malmberget and at the Ignalina plant, which are at the mercy of strong forces beyond the reach of the local community, including EU nuclear politics and the world market prices for iron ore. The ground on which people live is conditional, be it physically due to underground mining or economically because of uncertain prospects for the main employer and associated sources of income.

But the ground is shifting in Barsebäck and in Duisburg as well, in appearance and in significance. In Barsebäck, the nuclear plant is scheduled to disappear as completely as possible in favor of a reinstated "natural" landscape, and in Duisburg the ironworks as a place of production and work is gradually fading away as other activities and their material attributes take over the site. New meanings emerge from this instability, based on conceptualizations of the imperative to transform a toxic and dangerous environment into something friendly and livable. The new meanings also imply a focus on nature as a desirable feature in the new use—or the foreseen new use—of the sites, contributing to making them visible anew and eventually taking shape as something entirely different.

In the Ruhr district, former industrial sites are understood as parks and gardens with names like "Eden" and "Paradise" and even as industrial forests, with foresters employed to manage overgrown industrial land. At Barsebäck, the municipality envisions a future high-status seaside residential area on the cape where the nuclear power plant is still located, implying a scenario where children play in sandboxes and dogs are walked along the beachside promenades. In this vision, people are encouraged to invest in properties, trusting that the ground has been rid of radioactive contamination. To "clean and make green" was also the rallying cry of public opinion in Avesta for the transformation and reuse of the Koppardalen area.

Yet nature, be it as a conceptualization or as biotope, does not play the role of a straightforward redeemer in the process of healing post-industrial landscape wounds into scars. On the contrary, nature can be used to obscure dangers like radioactivity, and also to conceal past and present social injustices, like the romanticized pictures of decaying Detroit. This photographic genre is based on a dystopian but appealing framing that emphasizes the contrast between nature and the human-built world, but excludes people from the pictures, not to mention themes of social and economic inequalities or displacement. In many parts of the Western world, industrial ruins offer adventure and spectacle and speak to a contemporary longing for authenticity, forming the category of "ruined" post-industrial landscape scars. In some respects the industrial ruins represent the opposite of what the industrial enterprises previously stood for in terms of modernity, efficiency, welfare, and orderliness. In the post-industrial situation as ruins, they become an antithesis of their original function and character—a vanitas motif corresponding to the skulls and hourglasses of Flemish seventeenth-century paintings. Taken together, the recognition of a specific "industrial nature" pinpoints difficult issues of inclusion and exclusion of past and present experiences when a former industrial site is transformed.

The post-industrial landscape scars in this book show that different important stories about a place are carried by different groups of people. The stories might conflict with each other, they can be articulated in parallel, detached from each other. The scar has the potential to express manifold meanings in an integrated way, defined by such parameters as the logic of geographical scale and perceived distance. This is most apparent in the two cases dealing with nuclear power. The loyalty and trust expressed by those living close to a nuclear power plant, or working there, stand in striking contrast to the fear expressed by people living further away. The Barsebäck nuclear power plant was a despised silhouette when viewed from the Danish capital of Copenhagen across the Sound, but in the immediate vicinity the plant has practically merged with the visual landscape as its existence became a fact of everyday life. It was asserted by the employees at Barsebäck that when scared Danish people visited the nuclear power plant, they often lost most of their fear when they saw the plant from inside and experienced that there was nothing hidden or secret about it. Still, Barsebäck bears mythical overtones as the foremost symbol of nuclear power politics in Sweden and as such it continues to attract numerous visitors even since its closure.

The mythical appeal of nuclear power spans considerable distances. The Chernobyl accident in Soviet Ukraine not only caused radioactive contamination in parts of Sweden, the story of the event and the imagination of the postaccident situation hold a place in Swedish everyday metaphoric language. In Malmberget, far away from the nuclear business, youngsters talk about the Pit in the same breath as Chernobyl, both being described as deserted and repelling, albeit at different levels of danger. One small piece of the story of Chernobyl is thus found in its allusive prolongation at the edge of the Pit in Malmberget, carried by young people who were not even born when the accident occurred.

But is it always possible for the different stories and projections to coexist in peace, within a scar? Or will the wound continue to hurt and chafe? Whose scar is it eventually? In terms of national, ethnic, and social identities, the many stories of a scar can reveal deeply conflictual standpoints. For example, an articulation of the views and wishes of the inhabitants of the workers' town Sniečkus/Visaginas would differ substantially from the views and wishes held by the supporters of Sajūdis; the first group fighting for the dream of the Soviet paradise and the second fighting for the dream of independent Lithuania. To some extent, both dreams were challenged and betrayed.

The people of Barsebäck and Malmberget must also live with betrayed dreams. The vast investment in nuclear power, this "energy source of the future," the professional pride and hope for a better world turned into inflamed controversies and sharp lines between "us" and "them." In Malmberget, the squalor of the founding years was followed by a period when a welfare society actually materialized, only to be demolished in recent decades. In Avesta, the rediscovery of the abandoned industrial land did not really involve former workers, but was realized by newcomers, who focused on aesthetics and white-collar jobs.

Can the memory of the different dreams and the fights for these dreams be equally respected? Or is an acknowledgment of the scar merely a question of luxury? Is collective oblivion about difficult pasts not only understandable but also justifiable? Who could blame the antiquarians in Lithuania who had difficulties dealing with issues of heritage representing the Soviet period, and is it not liberating to find an alternative narrative in the Ruhr district, which puts the emphasis on a more manageable past—polluted nature—than on the moral depths of the two world wars? I hope this book has shown that it is indeed understandable, but not a question of luxury. Instead, what is critical is a search for a legitimate politics of memory. The scar does not need to be beautiful, but its meanings must be acknowledged.

The scar is also a possibility for reconstructing places of home, even though these places of home will not be exactly the same as before. Perhaps they will "persist by means of change," and, as such, challenge ideas about the meaning of concepts like stability and planning. In all five places, change stands out as a key characteristic. Change is decisive for the places not only in their transformation into post-industrial landscape scars, but also as an ongoing feature in how their significance is continuously negotiated. The Pit in Malmberget is perhaps the most striking example of how a place can change, both physically and mentally. The Pit forces successive but unpredictable abandonment of emotionally loaded places like one's home and the school of one's youth. One result is that the huge hole represents both presence—as a rumbling monster and an adventurous and forbidden territory—and absence—as the emptiness where the childhood places used to be and the good old days once happened. Affection for the physical environment in Malmberget becomes conditioned and to some extent repressed.

Continuous change takes other guises in Duisburg and in the Ruhr district in general. Here, change is both a conscious strategy and part of the attraction of the post-industrial landscape. From the perspective of industrial nature, ecologists suggest temporary biological conservation as a way to enhance biodiversity and offer recreational areas, and at the same time to make land available for new buildings. They foresee a map of mosaic cycles where pieces of land are set aside for about 15 years to become overgrown. After this period, the "wilderness" land is to be used for other purposes, and new land is set aside as replacement.

From the perspective of landscape architects, the continuous change of industrial nature in the Ruhr is a major element of the attraction for tourists and locals alike. In the densely built district, the former industrial areas intertwined with ruderal species and shrubs and even trees constitute a possibility for contemplating how the growing plants continually renew the spaces and to experience the shift of seasons. A capacity for process is a key focus for both ecologists and landscape architects engaging in industrial nature in the Ruhr.

The capacity for process also applies, I would argue, to larger questions of memory work. As was shown by examples of military facilities turned into wildlife reserves and former industrial sites transformed into commodified landscapes of consumption, industrial nature can also work as a visual rhetoric that conceals responsibilities for past and present social injustices. There is a thin line between processes of reconciliation and rekindling of hope for the future on the one hand, and processes of gentrification and the attraction of discovery on the other. A territory can simultaneously be understood as industrial *land*, meaning a place of work and of home, and as a post-industrial *landscape*, meaning a place of touristic gazes and visual desires. Intimate familiarity and distanced gazes do not simply represent good or bad ways of healing post-industrial landscape scars; their entanglement and complexity extend much further. I suggest that the scar constitutes a connection between the land and the landscape, turning the two understandings into simultaneous projections rather than chronologically successive definitions.

The ambiguity of the scar will remain, but its potential is that of a prospective approach. It is not just continuously added layers of meaning that define our memory work in the present, but an unfolding engagement that shapes new understandings, new places, and new worlds. The metaphor of the scar directs our glance toward the difficult and unstable places and stories, toward the closeness of the physical and the mental, and toward the temporally uneven and spatially widespread process of sharing past and present experiences.

Notes

1 Introduction

1. Anna Bohlin, "Land restitution and reconciliation: A lost opportunity? Emotion, land and heritage in post-apartheid South Africa," Paper presented at the healing-heritage workshop, University of Gothenburg, 2012. Cited in John Daniel Giblin, "Post-conflict heritage: Symbolic healing and cultural renewal," *International Journal of Heritage Studies* (2013), 13f.
2. See, for example, David Lowenthal, "Restoration: Synoptic reflections," in *Envisioning landscapes, making worlds: Geography and the humanities*, ed. Stephen Daniels (Milton Park, Abingdon, Oxon: Routledge, 2011), 209.
3. David Harvey, "Emerging landscapes of heritage," in *The Routledge companion to landscape studies*, ed. Peter Howard, Ian Thompson, and Emma Waterton (Abingdon: Routledge, 2013), 154f.
4. See, for example, Brian J. Graham and Peter Howard, eds., *The Ashgate research companion to heritage and identity* (Aldershot: Ashgate, 2008).
5. Shiloh R. Krupar, *Hot spotter's report: Military fables of toxic waste* (Minneapolis: University of Minnesota Press, 2013), 17.
6. See, for example, Timothy J. LeCain, *Mass destruction: The men and giant mines that wired America and scarred the planet* (New Brunswick, NJ: Rutgers University Press, 2009); Kenneth E. Foote, *Shadowed ground: America's landscapes of violence and tragedy*, revised edition (Austin: University of Texas Press, 2003), 23, 25, 337, 54, 57; Fredrik Krohn Andersson, *Kärnkraftverkets poetik: Begreppsliggörandet av svenska kärnkraftverk 1965–1973* (Stockholm: Stockholms universitet, diss., 2012), 104, 15; Michel Rautenberg, "Industrial heritage, regeneration of cities and public policies in the 1990s: Elements of a French/British comparison," *International Journal of Heritage Studies* 18, no. 5 (2012), 516; Magnus Rodell, "Monumentet på gränsen: Om den rumsliga vändningen och ett fredsmonument," *Scandia* 74, no. 2 (2008), 19.

7. Donna Houston, "Environmental justice storytelling: Angels and isotopes at Yucca Mountain, Nevada," *Antipode* 45, no. 2 (2013); Danielle Endres, "Sacred land or national sacrifice zone: The role of values in the Yucca Mountain participation process," *Environmental Communication* 6, no. 3 (2012); Foote, *Shadowed ground*, 25; Julia Fox, "Mountaintop removal in West Virginia: An environmental sacrifice zone," *Organization & Environment*, no. 12 (1999).
8. Utøya memorial, www.bustler.net/index.php/article/swedish _artist_jonas_dahlberg_to_design_july_22_memorial_sites_in_ norway/ (accessed April 9, 2014); see also the Vietnam Veterans Memorial by Maya Lin, inaugurated in 1982 and described as a scar. http://www.nybooks.com/articles/archives/2000/nov/02/making-the-memorial/?page=1 (accessed April 12, 2014).
9. Lebbeus Woods, *Radical reconstruction* (New York: Princeton Architectural Press, 1997); see also architectural work on war damaged buildings in Tina Wik, "Bosnia-Herzegovina, "Restoring war damaged built heritage in Bosnia-Herzegovina" (Conference report Bhopal, 2011). Available at http://tinawikarkitekter.se/publikationer/ (accessed April 9, 2014).
10. Woods, *Radical reconstruction*.
11. Ibid., 16.
12. Graham and Howard, *The Ashgate research companion*, 2; see also Laurajane Smith, *Uses of heritage* (New York: Routledge, 2006); Lisanne Gibson and John Pendlebury, eds., *Valuing historic environments* (Farnham: Ashgate, 2009), 1.
13. David Lowenthal, "Stewarding the future," *CRM: The Journal of Heritage Stewardship*, no. 2 (2005); Ola W. Jensen and Håkan Karlsson, *Archaeological conditions: Examples of epistemology and ontology*, GOTARC. Serie C, Arkeologiska skrifter, 0282–9479 (Göteborg: Göteborgs Universitet, 2000); Lowenthal, Jensen and Karlsson cited in Cornelius Holtorf and Anders Högberg, "Heritage futures and the future of heritage," in *Counterpoint: Essays in archaeology and heritage studies in honour of professor Kristian Kristiansen*, ed. Sophie Bergerbrant and Serena Sabatini (Oxford: Archaeopress, 2013), 741.
14. Hugh Cheape, Mary-Cate Garden, and Fiona McLean, "Editorial: Heritage and the environment," *International Journal of Heritage Studies* 15, no. 2–3 (2009), 105.
15. Sharon Macdonald, *Difficult heritage: Negotiating the Nazi past in Nuremberg and beyond* (London: Routledge, 2009); William Stewart Logan and Keir Reeves, eds., *Places of pain and shame: Dealing with "difficult heritage"* (Abingdon: Routledge, 2009); Dietrich Soyez, "Europeanizing industrial heritage in Europe: Addressing its transboundary and dark sides," *Geographische Zeitschrift* 91, no. 1 (2009); Giblin, "Post-conflict heritage;" see also, for example, Lars-Eric

Jönsson and Birgitta Svensson, eds., *I industrisamhällets slagskugga: Om problematiska kulturarv* (Stockholm: Carlsson, 2005); Eva Silvén and Anders Björklund, eds., *Svåra saker: Ting och berättelser som upprör och berör* (Stockholm: Nordiska museets förlag, 2006); Florence Fröhlig, *Painful legacy of World War II: Nazi forced enlistment. Alsatian/Mosellan prisoners of war and the Soviet prison camp of Tambov*, Stockholm Studies in Ethnology 8 (Stockholm: Acta Universitatis Stockholmiensis, diss., 2013); Beate Feldmann Eellend, *Visionära planer och vardagliga praktiker: Postmilitära landskap i Östersjöområdet*, Stockholm Studies in Ethnology 7 (Stockholm: Acta Universitatis Stockholmiensis, diss., 2013); Kyrre Kverndokk, *Pilegrim, turist og elev: Norske skoleturer til døds- og konsentrasjonsleirer* (Linköping: Linköpings Universitet, diss., 2007).
16. Giblin, "Post-conflict heritage," 4ff.
17. Ibid., 4.
18. See, for example, Jan af Geijerstam, *Landscapes of technology transfer: Swedish ironmakers in India 1860–1864*, Jernkontorets bergshistoriska skriftserie: 42. Stockholm papers in the history of philosophy of technology, Trita-HOT: 2045 (Stockholm: Jernkontoret, 2004), 11–16; see also chapter 4 in this book.
19. Smith, *Uses of heritage*, 44; see also Foote, *Shadowed ground*, 6.
20. Paul Ricœur, *Memory, history, forgetting*, trans. Kathleen Blamey and David Pellauer (Chicago: University of Chicago Press, 2004), 79, quote from 200; Doreen Massey, "Places and their pasts," *History Workshop Journal* 39 (1995); Mario Blaser, *Storytelling globalization from the Chaco and beyond* (Durham: Duke University Press, 2010); Shari Stone-Mediatore, *Reading across borders: Storytelling and knowledges of resistance* (New York: Palgrave Macmillan, 2003); Blaser and Stone-Mediatore cited in Houston, "Environmental justice storytelling"; for a related connection made between "landscape, memory and the pain of friction," see Sverker Sörlin, "Friction in the field: Meanings of military landscapes," in *Militære landskap: festspillutstillingen 2008 = Military landscapes: Bergen international festival exhibition 2008*, ed. Ingrid Book (Bergen: Bergen kunsthall, 2008), 43.
21. Amiria J. M. Henare, Martin Holbraad, and Sari Wastell, eds., *Thinking through things: Theorising artefacts ethnographically* (Milton Park, Abingdon, Oxon: Routledge, 2007), 3, 27.
22. Ibid., 7, 13, quote from 15; see also David E. Nye, *Technology matters: Questions to live with* (Cambridge, MA: MIT Press, 2006); Sven Widmalm and Hjalmar Fors, eds., *Artefakter: industrin, vetenskapen och de tekniska nätverken* (Hedemora: Gidlund, 2004).
23. See, for example, Daniel Bell, *The coming of post-industrial society: A venture in social forecasting* (Harmondsworth: Penguin, 1973); Bo Gustafsson, ed., *Post-industrial society: Proceedings of an international*

symposium held in Uppsala from 22 to 25 March 1977 to mark the occasion of the 500th anniversary of Uppsala university (London: Croom Helm, 1979); Krishan Kumar, *Prophecy and progress: The sociology of industrial and post-industrial society* (Harmondsworth: Penguin, 1978); Yoneji Masuda, *The information society: As postindustrial society* (Tokyo: Institute for the Information Society, 1980); Seymour Martin Lipset, ed., *The third century: America as a post-industrial society* (Stanford University: Hoover institution Press, 1980 [1979]); see also Maths Isacson, "Tre industriella revolutioner?," in *Industrialismens tid: Ekonomisk-historiska perspektiv på svensk industriell omvandling under 200 år*, ed. Maths Isacson and Mats Morell (Stockholm: SNS förlag, 2002); David Harvey, *The condition of postmodernity: An enquiry into the origins of cultural change* (Malden, MA: Blackwell Publishers Inc, 1990), vii.
24. Zygmunt Bauman, "From pilgrim to tourist—or a short history of identity," in *Questions of cultural identity*, ed. Stuart Hall and Paul du Gay (London: SAGE Publications Ltd, 1996), 29.
25. Marc Antrop, "A brief history of landscape research," in *The Routledge companion to landscape studies*, ed. Peter Howard, Ian Thompson, and Emma Waterton (Abingdon: Routledge, 2013), 12; Kenneth Robert Olwig, *Landscape, nature, and the body politic: From Britain's renaissance to America's new world* (Madison: University of Wisconsin Press, 2002); Rodell, "Monumentet på gränsen."
26. Mattias Qviström and Katarina Saltzman, "Exploring landscape dynamics at the edge of the city: Spatial plans and everyday places at the inner urban fringe of Malmö, Sweden," *Landscape Research* 31, no. 1 (2006), 22.
27. Yi-Fu Tuan, *Topophilia: A study of environmental perception, attitudes, and values* (New York: Columbia University Press, 1974), 132f.; Harvey, "Emerging landscapes of heritage," quote from 155.
28. John Urry, "The place of emotions within place," in *Emotional geographies*, ed. Joyce Davidson, Liz Bondi, and Mick Smith (Aldershot: Ashgate, 2005), 78f.; for a related analysis, understanding industrial sites as landscapes, see Svava Riesto, *Digging Carlsberg: Landscape biography of an industrial site undergoing redevelopment* (Forest & Landscape: University of Copenhagen, 2011).
29. Annika Alzén, *Fabriken som kulturarv: Frågan om industrilandskapets bevarande i Norrköping 1950–1985* (Stockholm/Stehag: Brutus Östlings Bokförlag Symposion, 1996), 13ff.
30. Marie Nisser, *Industriminnen: En bok om industri- och teknikhistoriska bebyggelsemiljöer* (Stockholm, 1979); Nisser cited in Alzén, *Fabriken som kulturarv*, 23.
31. Anders Houltz, *Teknikens tempel: Modernitet och industriarv på Göteborgsutställningen 1923*, Stockholm papers in the history and philosophy of technology, Trita-HOT: 2041 (Hedemora: Gidlund, 2003).

NOTES 163

32. Peter Aronsson, *Historiebruk: Att använda det förflutna* (Lund: Studentlitteratur, 2004), 286.
33. Iain J. M. Robertson, ed., *Heritage from below* (Farnham: Ashgate Publishing Company, 2012); R. Angus Buchanan, *Industrial archaeology in Britain* (Harmondsworth: Penguin Books Ltd, 1972), 19; Annika Alzén, "Kulturarv i rörelse: En jämförande studie," in *Kulturarvens gränser: Komparativa perspektiv*, ed. Peter Aronsson et al. (Göteborg: Bokförlaget Arkipelag, 2005), 219.
34. Bella Dicks, *Heritage, place, and community* (Cardiff: University of Wales Press, 2000); Marc Maure, "Identitet, økologi, deltakelse: Om museenes nye rolle," in *Økomuseumsboka: Identitet, økologi, deltakelse*, ed. John Aage Gjestrum and Marc Maure (Tromsø: Norsk ICOM, 1988), 24.
35. Michael Rix, "Industrial archaeology," *The Amateur Historian* 2, no. 8 (1955); Barrie Trinder, ed., *The Blackwell encyclopedia of industrial archaeology* (Oxford, 1992), 350; Kenneth Hudson, "Ecomuseums become more realistic," *Nordisk Museologi*, no. 2 (1996); Hugues de Varine, "L´Ecomusée," *Gazette*, no. 11, 2 (1978); De Varine related in John Aage Gjestrum, "En bibliografi om økomuseer," *Nordisk Museologi*, no. 2 (1996); Peter Davis, *Ecomuseums: A sense of place*, Leicester museum studies series (London: Leicester University Press, 1999), 67ff., 228; Kenneth Hudson, *Industrial archaeology: An introduction* (London: John Baker Publishers Ltd, 1963); Neil Cossons and Kenneth Hudson, eds., *Industrial archaeologists' guide 1969–70* (Newton Abbot Devon, 1969); Neil Cossons, *The BP book of industrial archaeology*, third ed. (Newton Abbot Devon: David & Charles, 1993); Eric N. Delony, "Industrial archeology in the United States 1981–1984," in *Industrial heritage '84 national reports: The fifth international conference on the conservation of the industrial heritage* (Washington DC, 1984); Anna Storm, *Hope and rust: Reinterpreting the industrial place in the late 20th century* (Stockholm: KTH diss., 2008); Gunnar Sillén, *Stiga vi mot ljuset: Om dokumentation av industri- och arbetarminnen* (Stockholm, 1977); Sven Lindqvist, *Gräv där du står: Hur man utforskar ett jobb* (Stockholm, 1978).
36. Marie Nisser, "Industriminnen under hundra år," *Nordisk Museologi*, no. 1 (1996).
37. Henrik Harnow, *Danmarks industrielle miljøer* (Odense: Syddansk Universitetsforlag, 2011, 11-56); Laurajane Smith, Paul Shackel, and Gary Campbell, eds., *Heritage, labour, and the working classes* (Abingdon, Oxon: Routledge, 2011), 2.
38. Smith, *Uses of heritage*, 29ff.; Anders Houltz, "Fabriken som aldrig blir färdig: Volvo Torslandaverken och 1960-talets svenska industrinationalism," *Fabrik og bolig*, no. 2012 (2012); Maja Fjæstad, "Ett kärnkraftverk återuppstår: Från SNR300 till Wunderland Kalkar," *Bebyggelsehistorisk tidskrift*, no. 63 (2012); Soyez, "Europeanizing

industrial heritage in Europe"; see also Geijerstam, *Landscapes of technology transfer*; Dag Avango, *Sveagruvan: Svensk gruvhantering mellan industri, diplomati och geovetenskap 1910–1934*, Stockholm papers in history and philosophy of technology, Trita-HOT: 2048 (Stockholm: Jernkontoret, 2005).
39. Foote, *Shadowed ground*, 25.
40. Michael Stratton, "Understanding the potential: Location, configuration and conversion options," in *Industrial buildings: Conservation and regeneration*, ed. Michael Stratton (London: E & FN Spon, 2000), 42ff.
41. Sharon Zukin, *Loft living: Culture and capital in urban change* (New Brunswick, NJ: Rutgers University Press, 1982), 59; Jussi S. Jauhiainen, "Waterfront redevelopment and urban policy: The case of Barcelona, Cardiff and Genoa," *European planning studies* 3, no. 1 (1995), 3.
42. Marie Nisser, "Industriminnen på den internationella arenan," in *Industrisamhällets kulturarv: Betänkande av Delegationen för industrisamhällets kulturarv* (Stockholm: Statens offentliga utredningar (SOU) 2002:67, 2002), 223; Betsy Bahr, "Adapting a future from the past: Reusing old industrial buildings for new industrial uses," in *Industrial heritage '84 proceedings: The fifth international conference on the conservation of the industrial heritage* (Washington DC, 1984); Bo Hedskog, *Återanvändning av industri- och specialbyggnader: Fastighetsekonomiska, tekniska och funktionella aspekter på val av ny användning*, TRITA-FAE-1012 meddelande 5:12 (Stockholm: Institutionen för fastighetsekonomi, KTH, 1982); *Sanering efter industrinedläggningar: Betänkande av industrisaneringsutredningen* (Stockholm: Statens offentliga utredningar (SOU) 1982:10, 1982).
43. Jauhiainen, "Waterfront redevelopment and urban policy," 3; Walter C. Kidney, *Working places: The adaptive use of industrial buildings* (Pittsburgh: Ober Park Associates, Inc., 1976), xf.
44. Delony, "Industrial archeology in the United States 1981–1984," 121.
45. JM, "Kvarnen: Warehouse living på Norra Älvstranden," Advertising brochure (2007), 16.
46. Ibid., 8.
47. "'Cultural' industrial plant engineer KRESTA: Stage setting for Bregenz Festival," *GAW group imteam: News from the group*, no. 1 (2005).
48. Jauhiainen, "Waterfront redevelopment and urban policy," 7, 20; Gene Desfor and John Jørgensen, "Flexible urban governance: The case of Copenhagen's recent waterfront development," *European planning studies* 12, no. 4 (2004), 489f.; Zukin, *Loft living*, 173.
49. Sharon Zukin, *Landscapes of power: From Detroit to Disney World* (Berkeley: University of California Press, 1991; repr., First Paperback Printing), 193.

50. Desfor and Jørgensen, "Flexible urban governance," 480; see also Håkon W. Andersen et al., *Fabrikken* (Oslo: Scandinavian Academic Press/Spartacus Forlag, 2004), 624.
51. Sverker Sörlin, "The trading zone between articulation and preservation: Production of meaning in landscape history and the problems of heritage decision-making," in *Rational decision-making in the preservation of cultural property: Report of the 86th Dahlem workshop, Berlin, March 26–31, 2000*, ed. Norbert S. Baer and Folke Snickars (Berlin: Dahlem University Press, 2001), 56; Zukin, *Loft living*, 68, 71.
52. *Sanering efter industrinedläggningar*. appendix 2:17, 4:13; Yttrande över industrisaneringsutredningens betänkande (SOU 1982:10) Sanering efter industrinedläggningar. Riksantikvarieämbetet och Statens historiska museer, Byrådirektör Staffan Nilsson, 2413/82, November 1, 1982. Marie Nisser's private collections.
53. Cossons, *The BP book of industrial archaeology*, 17ff.
54. David Lowenthal, *The past is a foreign country* (Cambridge: Cambridge University Press, 1985), 148–55.
55. Tim Edensor, *Industrial ruins: Spaces, aesthetics, and materiality* (Oxford: Berg, 2005), 3ff.
56. Ninjalicious, *Access all areas: A user's guide to the art of urban exploration* (Toronto: Infiltration, 2005); Robert Willim, *Industrial cool: Om postindustriella fabriker* (Lund: Lunds universitet, 2008).
57. See, for example, Jan Jörnmark, *Övergivna platser* (Lund: Historiska media, 2007); Edensor, *Industrial ruins*, 36ff.
58. Edensor, *Industrial ruins*, 14, 17, 122.
59. Nate Millington, "Post-industrial imaginaries: Nature, representation and ruin in Detroit, Michigan," *International Journal of Urban and Regional Research* 37, no. 1 (2013); George Steinmetz, "Detroit: A tale of two crises," *Environment and Planning D: Society and Space* 27, no. 5 (2009).
60. For a related analysis of post-military landscapes, see Sverker Sörlin, "Friction in the field: Meanings of military landscapes," in *Militære landskap: festspillutstillingen 2008 = Military landscapes: Bergen international festival exhibition 2008*, ed. Ingrid Book (Bergen: Bergen kunsthall, 2008); Gunilla Bandolin and Sverker Sörlin, *Laddade landskap – värdering och gestaltning av teknologiskt sublima platser*, R-07-14 (Stockholm: SKB, 2007).
61. Göran Greider, *När fabrikerna tystnar: Dikter* (Stockholm: Bonnier, 1995), 138. The quoted paragraph constitutes one part of a longer poem titled "När fabrikerna tystnar." My translation of the Swedish original:
Fabrikerna förblev dock förbjudna städer.
Barndomar förflöt i tecknet av ett mysterium.
De vuxna hade en bortvänd, oåtkomlig sida.
När fabrikerna tystnar blir de åter synliga.
Först nu är de främmande för oss.

2 Unstable Mountain

1. Quote from *Jernkontorets Annaler* in 1819, cited in Göran Littke, *Malmberget 1888–1963* (Malmberget: LKAB, 1963).
2. The background description is generally based on Gösta Forsström, *Gällivare kommun. Del 1. Malmberget. Malmbrytning och bebyggelse* (Luleå: Norrbottens Museum, 1973).
3. Gösta Forsström and Bo Strand, *Gällivare kommun. Del 2. Gällivare. Tätort och landsbygd* (Luleå: Norrbottens Museum, 1977), 54.
4. Forsström, *Malmberget*, 158.
5. Ibid., 164–76.
6. Jennie Sjöholm, *Heritagisation of built environments: A study of the urban transformation in Kiruna, Sweden* (Luleå University of Technology: Licentiate thesis, 2013), 9; see also Gudrun Andersson, *Kvinnor och män i Malmbergets kåkstad: Med kåkstaden i Kiruna som referenspunkt*, D-uppsats (Luleå Tekniska Universitet, 2004).
7. Staffan Hansson, "Malm, räls och elektricitet: Skapandet av ett teknologiskt megasystem i Norrbotten 1880–1920," in *Den konstruerade världen: Tekniska system i historiskt perspektiv*, ed. Pär Blomkvist and Arne Kaijser (Eslöv: Brutus Östlings bokförlag Symposion, 1998).
8. Sjöholm, *Heritagisation of built environments*, 17f.; Lasse Brunnström, *Kiruna—ett samhällsbygge i sekelskiftets Sverige* (Umeå: Umeå Universitet, diss., 1981); Thomas Nylund, "Kiruna—att planera för stadsflytt," *Plan*, no. 3 (2007), 20f.
9. Cyber City web encyclopedia: http://urbanhistory.historia.su.se/cybercity/stad/kiruna/historia.htm (accessed November 17, 2013).
10. Malmberget. Gruvan och samhället. LKAB, undated, 3. Gällivare municipality.
11. Maria-Pia Boëthius, *Heder och samvete: Sverige och andra världskriget* (Stockholm: Ordfront, 1999), 124ff.; see also Yvonne Hirdman, Jenny Björkman, and Urban Lundberg, *Sveriges historia. 1920–1965* (Stockholm: Norstedt, 2012), 337.
12. Bernt Löfström, "Västtyskland—vår största malmkund," *LKAB-tidningen*, no. 3 (1967), quote from 11; see also Håkon W. Andersen et al., *Fabrikken* (Oslo: Scandinavian Academic Press/Spartacus Forlag, 2004), 577.
13. *Gällivare generalplan* (Stockholm: Eglers stadsplanebyrå, 1967).
14. Cited in Forsström, *Malmberget*, 170.
15. Cited in Sven Ljunggren, "Undersökningen av kaptensmalmen under Malmbergets samhälle," *LKAB-tidningen*, no. 2 (1959), 8.
16. Ibid.
17. "Mysterierna i Malmberget," *LKAB-tidningen*, no. 1 (1962), 14.
18. Arne S. Lundberg, "Malmbergets framtid," *LKAB-tidningen*, no. 3 (1961), 3.

19. Fredrik Gustafsson and Pär Isling, *Gropen som svalde ett samhälle: Gropen i media 1956–1979*, C-uppsats (Luleå Tekniska Universitet, 2005), 22; *Gällivare generalplan*, quote from 1.
20. *Gällivare generalplan*, 6.
21. Ibid.; Gustafsson and Isling, *Gropen som svalde ett samhälle*, 21.
22. Lundberg, "Malmbergets framtid," 3.
23. Ibid.
24. *Gällivare generalplan*, 2.
25. Ibid., 1f.
26. Ibid., 68.
27. Adolf Henriksson, "Det moderna Malmberget," *LKAB-tidningen*, no. 1 (1963). My translation of "förbrukningssamhälle."
28. Gustafsson and Isling, *Gropen som svalde ett samhälle*, 23, 29.
29. Ibid., 23.
30. Ibid., 24, 31.
31. Maths Isacson, *Industrisamhället Sverige: Arbete, ideal och kulturarv* (Lund: Studentlitteratur, 2007), 28.
32. Ibid., 23, 145.
33. Carl-Erik Linné, "Kommun och industri i samarbete," *LKAB-tidningen*, no. 1 (1964), 20.
34. "LKAB 1.10.1957–30.9.1967," *LKAB-tidningen*, no. 3 (1967), 7.
35. Linné, "Kommun och industri i samarbete," 20.
36. Ibid., 22; Lennart Thelin, interview, Malmberget, June 13, 2007.
37. Since not all subcontractors are included, the number of employees is actually higher, especially during the latest period. See *Bergverksstatistik 2012* (Uppsala: SGU, 2013), 22ff.
38. Lennart Johansson, interview, Malmberget, June 13, 2007; *Fördjupad översiktsplan för tätorten Gällivare-Malmberget-Koskullskulle* (Gällivare kommun, 2003), 18; Fördjupad översiktsplan: Gällivare-Malmberget-Koskullskulle. Koncept II Samrådshandling 2006-12-18 118. Gällivare municipality.
39. Lennart Johansson, interview; Tommy Nyström, interview, Malmberget, June 15, 2007.
40. Thelin, interview.
41. Personal communication with LKAB, December 8, 2010; in Sweden, in 2010 there were about five thousand persons employed in the mining industry, three thousand of whom worked in iron ore mines. Not all subcontractors are included, which means that the actual number of employees is higher. See *Bergverksstatistik 2012*, 22ff.
42. Beate Feldmann, "'The Pit' and 'the ghetto'. On heritage, identity and generation," in *Malmberget. Structural change and cultural heritage processes. A case study*, ed. Birgitta Svensson and Ola Wetterberg (Stockholm: The Swedish National Heritage Board, 2009), 31, 33.
43. Nyström, interview.

44. Lennart Johansson, interview; Fördjupad översiktsplan: Gällivare-Malmberget-Koskullskulle. Koncept II Samrådshandling 2006–12–18, 5. Gällivare municipality.
45. Fördjupad översiktsplan: Gällivare-Malmberget-Koskullskulle. Koncept II Samrådshandling 2006–12–18, 24. Gällivare municipality; Ortsanalys Gällivare. Lägesrapport 2006–12–15 MAF Arkitektkontor & ÅF Infrastruktur, 2006, 46. Gällivare municipality.
46. *Fördjupad översiktsplan för tätorten Gällivare-Malmberget-Koskullskulle*, 78ff.; Ortsanalys Gällivare. Lägesrapport 2006–12–15. Gällivare municipality.
47. Nyström, interview; Lennart Johansson, interview; Feldmann, "'The Pit' and 'the ghetto,'" 39.
48. Nyström, interview; see also Lennart Johansson, interview.
49. Jan-Olof Hedström, "Hur kan det vara lagligt att flytta staden Kiruna för att ge plats åt gruvan?," *Plan*, no. 3 (2007), 30.
50. *Fördjupad översiktsplan för tätorten Gällivare-Malmberget-Koskullskulle*, 26, 49–50; Thelin, interview.
51. Nyström, interview.
52. See, for example, Hanna Gelotte, Eva Dahlström Rittsél, and Anna Ulfstrand, *Nästa hållplats Södertälje* (Stockholm: Länsstyrelsen i Stockholms län, 2006), 12f., 16.
53. Thelin, interview.
54. Fördjupad översiktsplan: Gällivare-Malmberget-Koskullskulle. Koncept II Samrådshandling 2006–12–18, 71. Gällivare municipality.
55. Ibid., 70; Nyström, interview.
56. Fördjupad översiktsplan: Gällivare-Malmberget-Koskullskulle. Koncept II Samrådshandling 2006–12–18, 5, 42. Gällivare municipality.
57. Ingrid Martins Holmberg, "The historicisation of Malmberget," in *Malmberget. Structural change and cultural heritage processes. A case study*, ed. Birgitta Svensson and Ola Wetterberg (Stockholm: The Swedish National Heritage Board, 2009); E. Johansson and M. Olsén, *Gällivares och Malmbergets födelse. En väggmålning av Erling Johansson i Centralskolan, Gällivare* (Kalix: Gellivare Sockens Hembygdsförening, undated).
58. Fördjupad översiktsplan för tätorten Gällivare-Malmberget—Koskullskulle. Samrådsyttrande 2002–04–08, Länsstyrelsen i Norrbottens län, 2002, 4. Gällivare municipality.
59. Förslag till Fördjupad översiktsplan för tätorten Gällivare-Malmberget-Koskullskulle. Samrådsredogörelse 2002–06 Gällivare kommun, 2002. Gällivare municipality.
60. Ibid., 5.
61. Fördjupad översiktsplan: Gällivare-Malmberget-Koskullskulle. Koncept II Samrådshandling 2006–12–18, 21, 56. Gällivare municipality; Thelin, interview.

NOTES

62. Tina Hedlund, Peter Popper, and Peter Ögren, "Miljörapport 2005. LKAB Malmberget. Yttre miljö. Publicerad 2006–03–27" (LKAB, 2006), 63.
63. Thelin, interview.
64. Fördjupad översiktsplan: Gällivare-Malmberget-Koskullskulle. Koncept II Samrådshandling 2006–12–18, 45. Gällivare municipality; Thelin, interview; Lennart Johansson, interview.
65. "Ett av husen sprack när det flyttades," *Norrländska Socialdemokraten* (September 4, 2007); "Snart börjar husen rulla i Malmberget," *Norrländska Socialdemokraten* (March 29, 2008); Nyström, interview; Lennart Johansson, interview.
66. Ulf Normark cited in Gabriella Olshammar, "Ruin landscape: A problem or history?" in *Malmberget. Structural change and cultural heritage processes. A case study*, ed. Birgitta Svensson and Ola Wetterberg (Stockholm: The Swedish National Heritage Board, 2009), 20.
67. Fördjupad översiktsplan: Gällivare-Malmberget-Koskullskulle. Koncept II Samrådshandling 2006–12–18, 45. Gällivare municipality.
68. Ibid., 21; Nyström, interview; Förslag till Fördjupad översiktsplan för tätorten Gällivare-Malmberget-Koskullskulle. Samrådsredogörelse 2002–06, 5. Gällivare municipality.
69. Feldmann, "'The Pit' and 'the ghetto'"; Olshammar, "Ruin landscape."
70. Sjöholm, *Heritagisation of built environments*, 21.
71. Ibid., 21–29.
72. Ibid., 26.
73. Nylund, "Kiruna—att planera för stadsflytt," 21f.
74. Sjöholm, *Heritagisation of built environments*, 26.
75. Ibid., 29, 31.
76. Nylund, "Kiruna—att planera för stadsflytt," 23; see also Sjöholm, *Heritagisation of built environments*, 8.
77. Hedström, "Hur kan det vara lagligt att flytta staden Kiruna för att ge plats åt gruvan?" 30.
78. Ibid., 31.
79. Sjöholm, *Heritagisation of built environments*, 26.
80. Nylund, "Kiruna—att planera för stadsflytt," 20f.; Anders Wilhelmson, "Nya Kiruna," *Plan*, no. 3 (2007), 38.
81. Sjöholm, *Heritagisation of built environments*, 26f., 30.
82. Ibid., 30.
83. Ibid., 31.
84. Krister Olsson and Marcus Adolphson, "Stadsstruktur, kulturvärden och identitet: Framtida flytt av Kiruna stad." Stockholm: KTH, 2008, 17.
85. Nylund, "Kiruna—att planera för stadsflytt," 24f.
86. Sjöholm, *Heritagisation of built environments*, 26.

87. Henrik Orrje, *Konsten att gestalta offentliga miljöer: Samverkan i tanke och handling* (Stockholm: Statens konstråd, 2013), 152–63.
88. Anna Storm and Krister Olsson, "The pit: Landscape scars as potential cultural tools," *International Journal of Heritage Studies* 19, no. 7 (2013).
89. Svante Lindqvist, *Än kan man köpa en Portello på Sporthallsfiket. Berättelser omkring Malmfälten* (Gällivare: Gellivare Sockens Hembygdsförening, 1995), quote from 120f; K. Johansson, "Vid kanten av Gropen," in *Norrbottens museum. Människors upplevelser av att bo i Malmberget i dag. Norrbottenprojektet. En nutidsdokumentation 1994–1995* (Norrbottens Museum, 1996), 7.
90. Feldmann, "'The Pit' and 'the ghetto,'" 36.
91. Ibid.; Ridderström cited in ibid., 34.
92. Olshammar, "Ruin landscape," 21ff.
93. Feldmann, "'The Pit' and 'the ghetto'," 36.
94. Nyström, interview.
95. Olshammar, "Ruin landscape," 20.
96. Ibid., 15f.
97. S. Lisinski, "Hela Malmberget skakas av sprängningarna i gruvan," *Dagens Nyheter* (February 3, 2007); A. Munck, "Gropen i Malmberget växer," *Dagens Nyheter* (November 3, 2008).
98. Hedlund et al., "Miljörapport 2005" 64; Fördjupad översiktsplan: Gällivare-Malmberget-Koskullskulle. Koncept II Samrådshandling 2006-12-18, 85. Gällivare municipality.
99. Hedlund et al., "Miljörapport 2005," 64.
100. Ibid., 84; Fördjupad översiktsplan: Gällivare-Malmberget-Koskullskulle. Koncept II Samrådshandling 2006-12-18, 85. Gällivare municipality.
101. Lennart Johansson, interview; Nyström, interview.
102. Information till samtliga fastighetsägare, LKAB, 2006, 2. Gällivare municipality.
103. Ibid.
104. Nyström, interview; see also Lindqvist, *Än kan man köpa en Portello på Sporthallsfiket*, 9.
105. Hedlund et al., "Miljörapport 2005." 63.
106. Information till samtliga fastighetsägare, LKAB, 2006, 6. Gällivare municipality; Utvidgning av gruvan i Malmberget. Till dig som bor på Elevhemsområdet. LKAB, 2006. Gällivare municipality.
107. Jennie Sjöholm and Kristina L. Nilsson, *Malmfältens kulturmiljöprocesser* (Luleå: Luleå Tekniska Universitet, 2011), 38ff.
108. Fördjupad översiktsplan: Gällivare-Malmberget-Koskullskulle. Koncept II Samrådshandling 2006-12-18, 86. Gällivare municipality; Olshammar, "Ruin landscape."
109. LKAB, *Årsredovisning* (Luleå: LKAB, 2006), 38.

NOTES   171

110. Wikström cited in Olshammar, "Ruin landscape," 22; *Fördjupad översiktsplan för tätorten Gällivare-Malmberget-Koskullskulle*, 36; Fördjupad översiktsplan: Gällivare-Malmberget-Koskullskulle. Koncept II Samrådshandling 2006-12-18, 89. Gällivare municipality.
111. Thelin, interview; see also "*Fördjupad översiktsplan för tätorten Gällivare-Malmberget-Koskullskulle*"; Nyström, interview.
112. Erik Holmstedt and Sverker Sörlin, *Inte längre mitt hem. Malmberget 1969-1978, 2007-2008* (Luleå: Black Island Books, 2008), 57; see also Laurajane Smith, Paul Shackel, and Gary Campbell, eds., *Heritage, labour, and the working classes* (Abingdon, Oxon: Routledge, 2011), 1.
113. Ridderström cited in Feldmann, "'The Pit' and 'the ghetto,'" 28.
114. Sjöholm, *Heritagisation of built environments*, 9.
115. Fördjupad översiktsplan: Gällivare-Malmberget-Koskullskulle. Koncept II Samrådshandling 2006-12-18, 31. Gällivare municipality.

3 Distance of Fear

1. "Sydkraft bygger atomkraftverk i Barsebäck för 350 miljoner kr," *Arbetet* (December 23, 1965); The company was originally named "Sydsvenska Kraftaktiebolaget," although the abbreviation "Sydkraft" was often used. In 1977 the name was formally changed to "Sydkraft."
2. Ibid; "Planerna på atomkraftverk innebär industriellt uppsving," *Sydsvenska Dagbladet* (December 24, 1965); Bo Malmsten, interview, Flemingsberg, October 7, 2010.
3. "Sydkraft bygger atomkraftverk i Barsebäck för 350 miljoner kr," *Arbetet* (December 23, 1965).
4. Malmsten, interview. Malmsten is quoting a national politician at the time.
5. Ibid.; Oscar Bjurling, *Sydkraft—samhälle: En berättelse i text och bild* (Malmö: Sydkraft, 1982), 250.
6. "Planerna på atomkraftverk innebär industriellt uppsving," *Sydsvenska Dagbladet* (December 24, 1965).
7. Ibid.; Bjurling, *Sydkraft—samhälle*, 217ff.
8. Cited in Bjurling, *Sydkraft—samhälle*, 217.
9. "Löddeköpinge kommun planerar för Sydkrafts atomkraftverk," *Skånska Dagbladet* (May 3, 1968).
10. "Planerna på atomkraftverk innebär industriellt uppsving," *Sydsvenska Dagbladet* (December 24, 1965); "150 bostäder i Barsebäckshamn," *Skånska Dagbladet* (March 8, 1968).
11. Fredrik Krohn Andersson, *Kärnkraftverkets poetik: Begreppsliggöranden av svenska kärnkraftverk 1965-1973* (Stockholm: Stockholms

universitet, diss., 2012), 121; Vattenbyggnadsbyrån, "Kraftöverföring från Barsebäcksverket." Kävlinge kommun. Kommunstyrelsen. Rapporter Utredningar. Energi. Spridda år. FC I: 6. Undated, 1. Kävlinge municipal archives.
12. "Protokoll fört vid sammanträde med Löddeköpinge kommunalfullmäktige måndagen den 28 oktober 1968, kl. 19.15 å Folkets Hus." Löddeköpinge kommun. Kommunalfullmäktige. Protokoll 1967–1971. AI: 3. 1968, § 111. Kävlinge municipal archives; "Protokoll fört vid sammanträde med byggnadsnämnden i Löddeköpinge den 17 oktober 1968, kl. 18.30 å kommunalkontoret i Löddeköpinge." Löddeköpinge kommun. Byggnadsnämnden. Protokoll. 1968. AI:5. 1968, § 181. Kävlinge municipal archives; "Sammanträdesprotokoll, Byggnadsnämnden, Löddeköpinge kommun, 21 augusti 1969." Löddeköpinge kommun. Byggnadsnämnden. Protokoll. Åren 1965–69 samt 1973. 1969, § 201, Bil. Kävlinge municipal archives; It might be worth noting that the town plan was exhibited during three weeks in the middle of the summer vacation period. See "Sammanträdesprotokoll, Byggnadsnämnden, Löddeköpinge kommun, 3 juli 1969." Löddeköpinge kommun. Byggnadsnämnden. Protokoll. Åren 1965–69 samt 1973. 1969, § 187. Kävlinge municipal archives; Malmsten, interview; "Planerna på atomkraftverk innebär industriellt uppsving," *Sydsvenska Dagbladet* (December 24, 1965); Annelie Sjölander-Lindqvist, Anna Bohlin, and Petra Adolfsson, *Delaktighetens landskap: Tillgänglighet och inflytande inom kulturarvssektorn* (Stockholm: Riksantikvarieämbetet, 2010), 27.
13. Stefan Lindström, "Implementing the welfare state: The emergence of Swedish atomic energy research policy," in *Center on the periphery: Historical aspects of 20th-century Swedish physics*, ed. Svante Lindqvist, Marika Hedin, and Thomas Kaiserfeld (Canton, MA: Science History Publications, 1993), 179ff.
14. Arne Kaijser, "Redirecting power: Swedish nuclear power policies in historical perspective," *Annual Review of Energy and the Environment* 17 (1992), 441ff.; Arne Kaijser, "From tile stoves to nuclear plants—the history of Swedish energy systems," in *Building sustainable energy systems: Swedish experiences*, ed. Semida Silveira (Stockholm: Svensk byggtjänst, 2001), 71ff.
15. Jonas Anshelm, *Bergsäkert eller våghalsigt? Frågan om kärnavfallets hantering i det offentliga samtalet i Sverige 1950–2002* (Lund: Arkiv, 2006), 182.
16. Jonas Anshelm, *Mellan frälsning och domedag: Om kärnkraftens politiska idéhistoria i Sverige 1945–1999* (Eslöv: Brutus Östlings Bokförlag Symposion, 2000); this period is also described in Åsa Moberg, *Ett extremt dyrt och livsfarligt sätt att värma vatten: En bok om kärnkraft* (Stockholm: Natur och kultur, 2014).
17. Kaijser, "Redirecting power," 445f.

NOTES 173

18. Ibid., 447, 49f.
19. "Planerna på atomkraftverk innebär industriellt uppsving," *Sydsvenska Dagbladet* (December 24, 1965); "Atomkraftverk blir turistattraktion," *Skånska Dagbladet* (July 29, 1971); there had been local protests caused by a fear of radioactive emissions in Sweden back in 1959, in connection with the planning of the Ågesta plant south of Stockholm. See Anshelm, *Bergsäkert eller våghalsigt?* 25, 48.
20. Anshelm, *Bergsäkert eller våghalsigt?* 13, 18ff., 31, 36. Anshelm is referring to SOU 1976: 30–31.
21. Ibid., 180.
22. Krohn Andersson, *Kärnkraftverkets poetik*, 61ff.
23. Ibid., 95, 194f.
24. Ibid., 98, 172.
25. Vattenbyggnadsbyrån, "Kraftöverföring från Barsebäcksverket," Kävlinge kommun. Kommunstyrelsen. Rapporter Utredningar. Energi. Spridda år. FC I: 6. Undated, 9. Kävlinge municipal archives; Krohn Andersson, *Kärnkraftverkets poetik*, 125ff.
26. Krohn Andersson, *Kärnkraftverkets poetik*, 118f., 124.
27. Ibid., 173; Henrik Borg and Helen Sannerstedt, *Barsebäcks kärnkraftverk. Rapport 2006:57* (Regionmuseet Kristianstad/Landsantikvarien i Skåne, 2006), 35; Vattenbyggnadsbyrån, "Kraftöverföring från Barsebäcksverket," Kävlinge kommun. Kommunstyrelsen. Rapporter Utredningar. Energi. Spridda år. FC I: 6. Undated, 13f. Kävlinge municipal archives; see also "Atomkraftverk blir turistattraktion," *Skånska Dagbladet* (July 29, 1971).
28. Krohn Andersson, *Kärnkraftverkets poetik*, 153f.
29. "Atomkraftverk blir turistattraktion," *Skånska Dagbladet* (July 29, 1971).
30. Krohn Andersson, *Kärnkraftverkets poetik*, 94, 176.
31. Borg and Sannerstedt, *Barsebäcks kärnkraftverk*, 39; "Atomkraftverk blir turistattraktion," *Skånska Dagbladet* (July 29, 1971).
32. Borg and Sannerstedt, *Barsebäcks kärnkraftverk*, 33; Niclas Grahn, *Barsebäcksverkets lokalisering och nedläggning: Hur förutsättningar och omständigheter för ett kärnkraftverk kan komma att förändras*, Arbetsrapporter. Kulturgeografiska institutionen nr. 710 (Uppsala: Uppsala Universitet, 2010), 21.
33. Arne Kaijser, "Trans-border integration of electricity and gas in the Nordic countries, 1915–1992," *Polhem*, no. 1 (1997), 6.
34. Bjurling, *Sydkraft—samhälle*, 22, 54ff.
35. Kaijser, "Trans-border integration of electricity and gas in the Nordic countries, 1915–1992," 13, 18; Bjurling, *Sydkraft—samhälle*, 170.
36. Henry Nielsen and Henrik Knudsen, "The troublesome life of peaceful atoms in Denmark," *History and technology* 26, no. 2 (2010), 96; Henrik Knudsen, *Risøs reaktorer: Registrering og dokumentering*

af bevaringsværdige genstande fra Forskningscenter Risøs reaktorfaciliteter med henblik på at bevare Danmarks nukleare kulturarv (Bjerringbro: Kulturarvsstyrelsen, 2006), 5, 7.
37. Oluf Danielsen, *Atomkraften under pres. Dansk debat om atomkraft 1974–85* (Roskilde: Roskilde Universitetsforlag, 2006), 49f.
38. Bjurling, *Sydkraft—samhälle*, 295.
39. "Sammanträdesprotokoll, Tekniska nämnden, Löddeköpinge kommun, 23 februari 1971." Löddeköpinge kommun. Tekniska nämnden. Protokoll. Åren 1962–71 samt 1973. 1971. Kävlinge municipal archives; "Sammanträdesprotokoll, Tekniska nämnden, Löddeköpinge kommun, 13 augusti 1973." Löddeköpinge kommun. Tekniska nämnden. Protokoll. Åren 1962–71 samt 1973. 1973. Kävlinge municipal archives; from 650 in 1960, the community's population increased to 4,600 by 1976.
40. Borg and Sannerstedt, *Barsebäcks kärnkraftverk*, 15f.; Fredrik Krohn Andersson, "Kärnkraftens arkitektur," in *Kärnkraft retro*, ed. Jan Garnert, *Dædalus, Tekniska museets årsbok* (Stockholm: Tekniska museet, 2008), 119f.
41. "Energiplanering i Kävlinge kommun. Alternativa värmeplaner för tätorterna Kävlinge, Furulund och Löddeköpinge. Utförd av kommunen och Sydkraft, januari 1982." Kävlinge kommun. Kommunstyrelsen. Rapporter Utredningar. Energi. Spridda år. FC I: 6. 1982, 1, 3. Kävlinge municipal archives.
42. Danielsen, *Atomkraften under pres*, 60, 71, 253; *Atomkraft?* no. 21–22 (1978), 18. Jørgen Steen Nielsen's private collections.
43. Malmsten, interview.
44. Maria Taranger, interview, Barsebäck, November 3, 2010; Borg and Sannerstedt, *Barsebäcks kärnkraftverk*, 44.
45. Taranger, interview; Sjölander-Lindqvist et al., *Delaktighetens landskap*, 28; the photo calendar was sent out starting in 1988.
46. "Sammanträdesprotokoll 1994–08–25." Kävlinge kommun. Lokala säkerhetsnämnden vid Barsebäcksverket. Protokoll. 1990–94. 1994, § 283. Kävlinge municipal archives; the shift in technology began in 1990 and was implemented over the following decade.
47. "Utredning om system för varning av allmänheten runt de svenska kärnkraftverken. Utvärding av erfarenheterna av RDS-systemet och förslag till ett mera flexibelt varningssystem med plan för genomförande och kostnadsberäkning. Räddningsverket 1995–08–11. Remissutgåva 2." Kävlinge kommun. Lokala säkerhetsnämnden. Ämnesordnade handlingar. 1995. F:4. 1995, 4. Kävlinge municipal archives; "Sammanträdesprotokoll 1982–02–22." Kävlinge kommun. Lokala säkerhetsnämnden vid Barsebäcksverket. Protokoll. 1981–89. AI:1. 1982, § 5. Kävlinge municipal archives.
48. Roland Palmqvist, interview, Kävlinge, October 31, 2011; Dalia Gineitiene et al., "Public risk perceptions of nuclear power. The

case of Sweden and Lithuania," in *Social processes and the environment*. *Lithuania and Sweden*, ed. Anna-Lisa Lindén and Leonardas Rinkevičius (Lund: Lund University, 1999), 136, 59; see also Jan Pålsson, cited in Borg and Sannerstedt, *Barsebäcks kärnkraftverk*, 64; Taranger, interview.
49. For a description of a relationship between the plant management and the local population marked by distrust, see Françoise Zonabend, *The nuclear peninsula*, translated by J. A. Underwood (Cambridge: Cambridge University Press 1993), 51.
50. Palmqvist, interview.
51. "Till ordföranden i Lokala säkerhetsnämnden vid Barsebäcksverket, 1991-04-24." Kävlinge kommun. Lokala säkerhetsnämnden. Korrespondens. 1991. E:1. 1991. Kävlinge municipal archives; "Lokalpolitiker riktar skarp kritik mot Barsebäcksverket: Brandsäkerheten är usel," *Sydsvenska Dagbladet* (May 6, 1991); "Föråldrat brandskydd ett hot mot Barsebäck. Lokal expert fruktar brand mer än härdsmälta," *Arbetet* (May 6, 1991).
52. "Beträffande brandskyddet vid Barsebäcksverket. SKI 1991-05-07." Kävlinge kommun. Lokala säkerhetsnämnden. Korrespondens. 1991. E:1. 1991. Kävlinge municipal archives; "SKI planerar brandsyn på alla kärnkraftverk," *Hallandsposten* (May 7, 1991); "Ang entledigandet av Jan Andersson i lokala säkerhetsnämnden." Kävlinge kommun. Lokala säkerhetsnämnden. Korrespondens. 1991. E:1. 1991. Kävlinge municipal archives.
53. Anshelm, *Mellan frälsning och domedag*.
54. Borg and Sannerstedt, *Barsebäcks kärnkraftverk*, 59.
55. Bjurling, *Sydkraft—samhälle*, 251.
56. Kaijser, "Redirecting power," 452f.
57. Ibid., 453.
58. Anshelm, *Mellan frälsning och domedag*.
59. Caj Andersson, "Om vi kände skräck för kärnkraften skulle vi inte arbeta här!," *Året runt* 46 (1979).
60. Ibid.
61. Borg and Sannerstedt, *Barsebäcks kärnkraftverk*, 21.
62. Kaijser, "Redirecting power," 454; see also Maja Fjæstad, "Demokratins triumf eller fiasko? Folkomröstningen om kärnkraft i retrospektiv," in *Kärnkraft retro*, edited by Jan Garnert. Dædalus, Tekniska museets årsbok, 65–75 (Stockholm: Tekniska museet, 2008).
63. Anshelm, *Mellan frälsning och domedag*.
64. This period is described in Åsa Moberg, *Ett extremt dyrt och livsfarligt sätt att värma vatten: En bok om kärnkraft* (Stockholm: Natur och kultur, 2014); Lennart Daléus, interview, Saltsjöbaden, October 5, 2010.
65. Leif Öst, interview, Barsebäck, November 3, 2010.

66. Ibid.; Anshelm, *Bergsäkert eller våghalsigt?* 50f.; Bjurling, *Sydkraft—samhälle*, 249, 79f.
67. Bjurling, *Sydkraft—samhälle*, 281.
68. Kaijser, "Redirecting power," 455.
69. Bjurling, *Sydkraft—samhälle*, 281.
70. Danielsen, *Atomkraften under pres*, 79.
71. *Atomkraft?* no. 1 (1974), 5. Jørgen Steen Nielsen's private collections; *En fremtid med atomkraft? Organisationen til oplysning om atomkraft* (1975). Jørgen Steen Nielsen's private collections.
72. Danielsen, *Atomkraften under pres*, 259, 99, 314ff., 28.
73. Bent Sørensen, "Rasmussen rapporten skaber tvivl. Er Gyllingnæs sikker?" *Atomkraft?* no. 15 (1976). Jørgen Steen Nielsen's private collections; Danielsen, *Atomkraften under pres*, 65f.; Jørgen Steen Nielsen, interview, Köpenhamn, November 2, 2010; *En alternativ energiplan for Danmark*, (København: OOA, 1983).
74. Danielsen, *Atomkraften under pres*, 15.
75. Ibid., 12, 658.
76. Jørgen Steen Nielsen, "Overblik," *Atomkraft?* no. 25–26 (1979). Jørgen Steen Nielsen's private collections; Knudsen, *Risøs reaktorer*, 100.
77. Steen Nielsen, interview.
78. Siegfried Christiansen, "Disse selskaber og deres prægtige atomkraftværker," *Politiken*, (November 30, 1973); Arne Schiøtz, "Vort ansvar over for vore efterkommere," *Politiken*, (February 6, 1974); Jørgen Steen Nielsen, "Vi må ikke handle i blinde," *Berlingske tidene* (February 3, 1974); Jørgen Boldt, "Atomkraftens pris," *Information*, (January 19–20, 1974); Siegfried Christiansen, "Atomkraftens skygge sider," *Extra bladet*, (January 26, 1974); Jørgen Boldt, "Hjemmelavet atombombe," *Politiken*, (January 20, 1974); Siegfried Christiansen, "ELSAMs kamp for at få atomkraftværkerne ind ad den danske bagdør," *Information*, (February 6, 1974); *"aldrig gi'r vi op..." Atomkraftmarcherne 1978*, København: OOA, 1978. Jørgen Steen Nielsen's private collections; Danielsen, *Atomkraften under pres*, 702; Steen Nielsen, interview.
79. "Letter from Anne Lund" (April 4, 1975). D7a, 1321/2001. National Museum of Denmark archives; Badge (1975). D16f, 692/2001. National Museum of Denmark archives; Mette-Line Thorup, "Solmærket, ja tak," *Information*, (July 25, 2007).
80. Thorup, "Solmærket, ja tak," *Information*, (July 25, 2007).
81. Steen Nielsen, interview; as in Denmark, antinuclear publications in Sweden were remarkably technical in their approach. See Anshelm, *Mellan frälsning och domedag*.
82. *Atomkraft?* no. 25–26 (1979), 2. Jørgen Steen Nielsen's private collections; Danielsen, *Atomkraften under pres*, 700f.; Steen Nielsen, interview.
83. "Inhold," *Atomkraft?* no. 25–26 (1979), 2. Jørgen Steen Nielsen's private collections; Danielsen, *Atomkraften under pres*, 255f.; Kære

Anker Jørgensen, "De 3 hovedkrav," *Atomkraft?* no. 25-26 (1979), 6. Jørgen Steen Nielsen's private collections.
84. Annette Wiborg, "Der har altid været tryk på..." *Atomkraft? OOA—ti år i bevægelse* no. 45 (1984), 24. Jørgen Steen Nielsen's private collections; Steen Nielsen, interview; "Kræver beslutning om atomværker i Danmark udskudt 3 år," *Ekstra bladet* (1974); "En OOA-aktivist. Beretning fra Helle Green," 165/99 (2000). National Museum of Denmark archives; "OOA memories, written by P.J. Snare" 165/99 (2000). National Museum of Denmark archives; Danielsen, *Atomkraften under pres*, 698.
85. Energibevægelsen OOA. Organisationen til Oplysning om Atomkraft, www.ooa.dk (accessed June 19, 2012).
86. En fremtid med atomkraft? *Organisationen til oplysning om atomkraft* (1975), 22. Jørgen Steen Nielsen's private collections; Danielsen, *Atomkraften under pres*, 702.
87. *Atomkraft?* no. 25-26 (1979), 10. Jørgen Steen Nielsen's private collections.
88. Ibid., 14.
89. *Atomkraft?* no. 21-22 (1978), 23. Jørgen Steen Nielsen's private collections.
90. Daléus, interview; Steen Nielsen, interview.
91. See, for example, Steen Nielsen, "Overblik," *Atomkraft?* no. 25-26 (1979), 3. Jørgen Steen Nielsen's private collections; Danielsen, *Atomkraften under pres*, 178.
92. *Atomkraft? OOA—ti år i bevægelse* no. 45 (1984). Jørgen Steen Nielsen's private collections; *Atomkraft?* no. 1 (1974), cover page. Jørgen Steen Nielsen's private collections; "Myten om atomkraftens 'fredelige' udnyttelse," *En fremtid med atomkraft? Organisationen til oplysning om atomkraft* (1975), 14. Jørgen Steen Nielsen's private collections.
93. Per Högselius and Arne Kaijser, *När folkhemselen blev internationell: Elavregleringen i historiskt perspektiv* (Stockholm: SNS Förlag, 2007), 58.
94. Kaijser, "Redirecting power," 457.
95. Anshelm, *Mellan frälsning och domedag*, 352.
96. Ibid., 362f., 462; see also Ernst Klein, "Vi behöver kärnkraften," *Östgöta Correspondenten* (undated).
97. Danielsen, *Atomkraften under pres*, 549.
98. Palmqvist, interview; Daléus, interview.
99. Högselius and Kaijser, *När folkhemselen blev internationell*, 59f., quote from 60.
100. Borg and Sannerstedt, *Barsebäcks kärnkraftverk*, 70.
101. "1988-06-30 Inbjudan till information om den fortsatta verksamheten vid Barsebäcksverket." Kävlinge kommun. Kommunstyrelsen. Handlingar. Korrespondens. 1988. E I: b 6. 1988. Kävlinge municipal archives; "Kolförsök försenas?" *Norra Skåne* (January 13, 1989).

102. "Protokoll fört vid KSO-möte i Bad Nenndorf 1989-08-18." Kävlinge kommun. Lokala säkerhetsnämnden. Ämnesordnade handlingar. 1989. F:2. 1989, § 3. Kävlinge municipal archives.
103. Jan Thorsson, "Barsebäcksborna till kamp mot kolet," *Svenska Dagbladet* (January 23, 1989); Jan Jönsson, "Dahl kommer inte till möte om kolgasen," *Sydsvenska Dagbladet* (January 12, 1989); Sjölander-Lindqvist et al., *Delaktighetens landskap*, 32.
104. "Barsebäcksbor vill stoppa kolgasverk," *Kristianstadsbladet* (January 9, 1989).
105. Jan Thorsson, "Politkerna våndas," *Svenska Dagbladet* (January 23, 1989); "Letter from Miljö- och naturresursdepartementet to Sydkraft AB." Kävlinge kommun. Kommunstyrelsen. Handlingar. M-Ö. 1992. EIIc: 2. 1988. Kävlinge municipal archives.
106. Högselius and Kaijser, *När folkhemselen blev internationell*, 60f.
107. Danielsen, *Atomkraften under pres*, 387.
108. Quote in original: "Tack, ni svenska vakttorn. Med plutonium tvingar vi dansken på knä. Här: Danmark, utskitet av kalk och vatten. Och där: Sverige, hugget i granit. Danskjävlar. Danskjävlar!"
109. "Sydkraft klarar vintern med el från Danmark," *Nyheterna* (December 23, 1992); "Dansk el räddar skånska julen," *Sydsvenska Dagbladet* (December 23, 1992).
110. Thorsten Jansson, "Dansk kol-el håller Sverige ljust i jul," *Östra Småland* (December 23, 1992); Sven Bergquist, "Vad hände i Barsebäck den 28 juli?" *Expressen* (December 4, 1992).
111. Larry Andow, "Barsebäck—Danmarks alibi," *Arbetet* (September 9, 1982).
112. Pär Adolfsson, cited in Borg and Sannerstedt, *Barsebäcks kärnkraftverk*, 72.
113. Öst, interview.
114. Taranger, interview.
115. Sonje Johansson, cited in Borg and Sannerstedt, *Barsebäcks kärnkraftverk*, 66.
116. Ibid., 44.
117. Ibid., 48.
118. Ibid., 44.
119. Öst, interview.
120. Borg and Sannerstedt, *Barsebäcks kärnkraftverk*, 48; Öst, interview.
121. Borg and Sannerstedt, *Barsebäcks kärnkraftverk*, 49f.
122. Ibid., 61, 67.
123. Arne Hansson, cited in ibid., 62.
124. Arne Hansson, cited in ibid., 61f; see also Sjölander-Lindqvist et al., *Delaktighetens landskap*, 39.
125. Taranger, interview; see also Öst, interview.
126. Bjurling, *Sydkraft—samhälle*, 275.

NOTES 179

127. Taranger, interview; Öst, interview.
128. Leif Öst, cited in Borg and Sannerstedt, *Barsebäcks kärnkraftverk*, 47.
129. Taranger, interview.
130. Jan Pålsson, cited in Borg and Sannerstedt, *Barsebäcks kärnkraftverk*, 64; for a similar story told at the Clab storage in Oskarshamn, see Gunilla Bandolin and Sverker Sörlin, *Laddade landskap – värdering och gestaltning av teknologiskt sublima platser, R-07–14* (Stockholm: SKB, 2007), 38.
131. Zonabend, *The nuclear peninsula*. See, for example, 6, 101, 20.
132. Ibid., 93f.
133. In addition to previous references, Gineitiene et al. report on an opinion survey of the 1990s that includes Barsebäck. See Gineitiene et al., "Public risk perceptions of nuclear power," 136, 59.
134. Jan Pålsson, cited in Borg and Sannerstedt, *Barsebäcks kärnkraftverk*, 66; Öst, interview.
135. Öst, interview.
136. Ibid.
137. Lars Tornsberg, cited in Borg and Sannerstedt, *Barsebäcks kärnkraftverk*, 69.
138. Sonje Johansson, cited in ibid., 71.
139. Taranger, interview.
140. Palmqvist, interview.
141. Åsa Carlson, interview, Malmö, November 4, 2010.
142. Energibevægelsen OOA. Organisationen til Oplysning om Atomkraft, www.ooa.dk (accessed June 19, 2012).
143. Birgitte Thusing, "Barsebäck værste terror-mål," *Extra bladet* (September 18, 2001).
144. Jens Aagaard Poulsen, "Vi er forsvarsløse," *Extra bladet* (September 18, 2001).
145. Dan Nilsson, "Höjd beredskap vid svenska kärnkraftverk," *Svenska Dagbladet* (September 18, 2001).
146. *Atomkraft? OOA—ti år i bevægelse* no. 45 (1984), 6. Jørgen Steen Nielsen's private collections; Danielsen, *Atomkraften under pres*: 5f., 187, 429; Energibevægelsen OOA. Organisationen til Oplysning om Atomkraft, www.ooa.dk (accessed June 19, 2012).
147. Palmqvist, interview.
148. Philip Moding, "Underlag för gemensamt PM till Energikomissionen—Kävlinge kommun, 1995–08-28, KK6226." Kävlinge kommun. Lokala säkerhetsnämnden. Ämnesordnade handlingar. 1995. F:4. 1995. Kävlinge municipal archives.
149. Carlson, interview.
150. Öst, interview.
151. Ibid.
152. Taranger, interview.

153. Sjölander-Lindqvist et al., *Delaktighetens landskap*, 35.
154. Kristian Berg and Lars-Eric Jönsson, "Öppet brev till länsantikvarien i Malmöhus län," *Helsingborgs Dagblad* (June 9, 1988).
155. Regionmuseet, *Stort, fult, farligt? Barsebäcks kärnkraftverk och kulturarvet. Rapport 2003:115* (Regionmuseet Kristianstad/Landsantikvarien i Skåne, 2003); Borg and Sannerstedt, *Barsebäcks kärnkraftverk*.
156. Magdalena Tafvelin Heldner, Eva Dahlström Rittsél, and Per Lundgren, *Ågesta: Kärnkraft som kulturarv* (Stockholm: Tekniska museet, Stockholms läns museum, Länsstyrelsen i Stockholms län, 2008); Jan Garnert, ed. *Kärnkraft retro* (Stockholm: Tekniska museet, 2008); at the National Museum of Science and Technology in Stockholm no fewer than eight exhibitions on nuclear power were shown between 1949 and 1999, mainly from an uncritical and supportive perspective. See Magdalena Tafvelin Heldner, "Strumpstickor och pingpongbollar: Med Tekniska museet i atomåldern," in *Kärnkraft retro*, ed. Jan Garnert, *Dædalus, Tekniska museets årsbok* (Stockholm: Tekniska museet, 2008).
157. During the planning before the seminar several of the heritage representatives did not find it self-evident to connect the nuclear power plant to heritage issues. See preface by Barbro Mellander in Borg and Sannerstedt, *Barsebäcks kärnkraftverk*, 9; Sjölander-Lindqvist et al., *Delaktighetens landskap*, 25.
158. Preface by Barbro Mellander in Borg and Sannerstedt, *Barsebäcks kärnkraftverk*, 9. See also 85, 92; Berg and Jönsson, "Öppet brev till länsantikvarien i Malmöhus län," *Helsingborgs Dagblad* (June 9, 1988).
159. "Letter from the National Museum to OOA activists," 165/99 (June 28, 2000). National Museum of Denmark archives; Energibevægelsen OOA. Organisationen til Oplysning om Atomkraft, www.ooa.dk (accessed June 19, 2012).
160. "En OOA-aktivist. Beretning fra Helle Green," 165/99 (2000). National Museum of Denmark archives; see also "OOA memories, written by P.J. Snare," 165/99 (2000). National Museum of Denmark archives.
161. "Sculpture 'Sigyn'" J2v, D7a, 1562/2001. National Museum of Denmark archives.
162. Visit to the museum, November 2, 2010.
163. Knudsen, *Risøs reaktorer*, 1, 7.
164. Ibid., 111.
165. Johan Joelsson, "Kärnkraft som attraktion," *DIK-forum*, no. 6 (2010).
166. Taranger, interview.
167. Borg and Sannerstedt, *Barsebäcks kärnkraftverk*, 100.
168. Jan Olof Andersson Hederoth, cited in ibid., 78; see also Sjölander-Lindqvist et al., *Delaktighetens landskap*, 37.

NOTES 181

169. Håkan Lorentz, interview, Barsebäck, November 3, 2010; Carlson, interview.
170. Sjölander-Lindqvist et al., *Delaktighetens landskap*: 30f.; these plans have recently met with resistance from the County Administrative Board. See Ulf Sundberg, "Länsstyrelsen sågar Barsebäcks sjöstad," *Sydsvenska Dagbladet* (April 14, 2014).
171. Sjölander-Lindqvist et al., *Delaktighetens landskap*, 30.
172. Henrik Borg, "Barsebäck, en historia om trivsel, säkerhet, demonstrationer och kulturarv," in *Kärnkraft retro*, ed. Jan Garnert, *Dædalus, Tekniska museets årsbok* (Stockholm: Tekniska museet, 2008), 143.

4 Lost Utopia

1. Paul Josephson, *Red Atom: Russia's nuclear power program from Stalin to today* (New York: W. H. Freeman and Company, 1999), 34.
2. Ibid., 35, 234.
3. See, for example, Sonja D. Schmid, "Celebrating tomorrow today: The peaceful atom on display in the Soviet Union," *Social Studies of Science* 36, no. 3 (2006), 356.
4. Arne Kaijser and Per Högselius, *Resurs eller avfall? Politiken kring hanteringen av använt kärnbränsle i Finland, Tyskland, Ryssland och Japan*, R-07–37 (Stockholm: SKB, 2007), 41; Andis Cinis, Marija Drėmaitė, and Mart Kalm, "Perfect representations of Soviet planned space: Mono-industrial towns in the Soviet Baltic Republics in the 1950s–1980s," *Scandinavian Journal of History* 33, no. 3 (2008), 237; Per Högselius, "Connecting East and West? Electricity systems in the Baltic region," in *Networking Europe. Transnational infrastructures and the shaping of Europe 1850–2000*, ed. Erik van der Vleuten and Arne Kaijser (Sagamore Beach, MA: Science History Publications, 2006), 251.
5. Cinis et al., "Perfect representations of Soviet planned space," 237; Kristina Šliavaitė, *From pioneers to target group. Social change, ethnicity and memory in a Lithuanian nuclear power plant community* (Lund: Lunds universitet, 2005), 64, 70f.
6. Šliavaitė, *From pioneers to target group*, 69f.; see also Kate Brown, "In the house that plutonium built: The history of plutonium, radiation and the communities that learned to love their bomb," 26. Unpublished.
7. Cited in Šliavaitė, *From pioneers to target group*, 70f.; combining Soviet and American contexts, Kate Brown describes how places were regarded as empty of history and then given a beginning and thus meaning through the establishment of a nuclear enterprise. See Kate Brown, "Gridded lives: Why Kazakhstan and Montana are nearly the same place," *The American Historical Review* 106, no. 1 (2001), 28.

8. Cited in Šliavaitė, *From pioneers to target group*, 65; see also Cinis et al., "Perfect representations of Soviet planned space," 241.
9. There are other terms used such as "atomograds," "mono-cities," and "company towns." Nevertheless, the best expression is perhaps "mono-industrial towns" since it was used in the Soviet Union and allows a distinction from Western-style company towns where the private patriarchal owner was the driving force for the establishment. See, for example, Cinis et al., "Perfect representations of Soviet planned space," 227f., 236f.; Nicole Foss, *Nuclear safety and international governance: Russia and Eastern Europe* (Oxford: Oxford Institute for Energy Studies, 1999), 76, note 109.
10. Anna Veronika Wendland, "Inventing the Atomograd. Nuclear as a way of life in Eastern Europe before and after Chernobyl" 8–9. Unpublished; Šliavaitė, *From pioneers to target group*, 70ff.
11. Cinis et al., "Perfect representations of Soviet planned space," 228, 239, quote from 239; for a similar characterization (Eden), see Brown, "In the house that plutonium built."
12. Šliavaitė, *From pioneers to target group*, 22f.
13. Ibid., 23, 45; Viktor Ševaldin, interview, Ignalina, November 25, 2010; the residents of secret cities have also been referred to as "chocolate eaters." See Brown, "In the house that plutonium built," 29; the material benefits were even more pronounced in plutonium-producing cities, "plutopias" to use Kate Brown's term. See Kate Brown, *Plutopia: Nuclear families, atomic cities, and the great Soviet and American plutonium disasters* (Oxford: Oxford University Press, 2013); The utopian character of secret cities encompassed not only the Soviet Union and the United States. For an example from Australia, see Gunilla Bandolin and Sverker Sörlin, *Laddade landskap – värdering och gestaltning av teknologiskt sublima platser, R-07-14* (Stockholm: SKB, 2007), 11.
14. Ševaldin, interview.
15. Šliavaitė, *From pioneers to target group*, 28, 43.
16. Cinis et al., "Perfect representations of Soviet planned space," 243, note 37; from the early decades of the Soviet Union, Soviet architects preferred to build new cities on virgin soil. See Brown, "Gridded lives," 26f.
17. Wendland, "Inventing the atomograd," 11; Cinis et al., "Perfect representations of Soviet planned space," 227.
18. Šliavaitė, *From pioneers to target group*, 22f., with reference to Algirdas Kavaliauskas, *Visaginas: Istorijos fragmentai (1972–2002)* (Vilnius: Jandrija, 2003), 47–48; another explanation could be that locals were not considered to be suitable inhabitants of this paradise. See a similar analysis in Brown, "In the house that plutonium built."
19. Liutauras Labanauskas, "Social aspects of the functioning of the Ignalina nuclear power plant," *Viešoji politika ir administravimas*, no. 22 (2007), 79f.

20. Many such new names of nuclear cities were derived from landscapes and rivers in the surroundings, from Soviet neologisms in the Russian imperial tradition or from Soviet heroes. See Wendland, "Inventing the Atomograd," 8–9.
21. In Sniečkus the new fashion of Soviet planners to take advantage of the natural landscape was elaborated, in sharp contrast to the Stalinist gridded cities of previous decades. See Cinis et al., "Perfect representations of Soviet planned space," 237ff., quote from 240; see also Brown, "Gridded lives."
22. Personal communication with a guide at the Ignalina nuclear power plant, November 25, 2010.
23. Kaijser and Högselius, *Resurs eller avfall?*, 41; a related argument is that the United States' military nuclear activity after World War II became a key nationbuilding project and a national fetish. See Joseph Masco, *The nuclear borderlands: The Manhattan project in post–Cold War New Mexico* (Princeton, NJ: Princeton University Press, 2006).
24. Kaijser and Högselius, *Resurs eller avfall?*, 42; Michael Herttrich, Rolf Janke, and Peter Kelm, "International co-operation to promote nuclear-reactor safety in the former USSR and Eastern Europe," in *Green Globe Yearbook of International Co-Operation on Environment and Development 1994*, ed. Helge Ole Bergesen and Georg Parmann (Oxford: Oxford University Press, 1994), 89.
25. Josephson, *Red Atom*, 281; Paul Josephson, "Technological utopianism in the twenty-first century. Russia's nuclear future," *History and technology* 19, no. 3 (2003), 282; Sovietologist blog, http://sovietologist.blogspot.fr/2008/04/rbmk-reactors-and-weapons-grade.html (accessed April 15, 2014).
26. Herttrich et al., "International co-operation to promote nuclear-reactor safety," 92; Gunnar Johanson, interview, Stockholm, August 14, 2012.
27. Kaijser and Högselius, *Resurs eller avfall?*, 42; Herttrich et al., "International co-operation to promote nuclear-reactor safety," 89ff.; Jane I. Dawson, *Eco-nationalism. Anti-nuclear activism and national identity in Russia, Lithuania, and Ukraine* (Durham: Duke University Press, 1996), 3.
28. Foss, *Nuclear safety and international governance*, 58; Herttrich et al., "International co-operation to promote nuclear-reactor safety," 93.
29. Josephson, *Red Atom*, 204; "Technological utopianism," 277; Foss, *Nuclear safety and international governance*, 7; energy, and especially nuclear energy involving uranium mining, has long been key feature of colonial and postcolonial relationships in many parts of the world. See Gabrielle Hecht, *Being nuclear: Africans and the global uranium trade* (Cambridge, MA: MIT Press, 2012); Gabrielle Hecht, ed., *Entangled geographies: Empire and technopolitics in the global Cold War* (Cambridge, MA: MIT Press, 2011).

30. Högselius, "Connecting East and West?" 245.
31. Ibid., 247f.
32. Ibid., 251.
33. Josephson, *Red atom*, 34.
34. Foss, *Nuclear safety and international governance*, 8.
35. Josephson, *Red atom*, 204; "Technological utopianism," 277.
36. Alfred Erich Senn, *Gorbachev's failure in Lithuania* (New York: St. Martin's Press, 1995), xiv; for the use of religious symbols as a mark against Soviet atheism, from 1989 and onward, see Tatiana Kasperski, "Chernobyl's aftermath in political symbols, monuments and rituals: Remembering the disaster in Belarus," *Anthropology of East Europe Review* 30, no. 1 (2012), 86.
37. Saulius Urbonavičius, interview, Ignalina, November 25, 2010.
38. Josephson, *Red atom*, 241f.
39. Susanne Bauer, Karena Kalmbach, and Tatiana Kasperski, "From Pripyat to Paris, from grassroots memories to globalized knowledge production: The politics of Chernobyl fallout," in *Nuclear Portraits*, ed. Laurel MacDowell (Toronto: University of Toronto Press, forthcoming); Wendland, "Inventing the atomograd," 13ff.; Josephson, *Red atom*, 244, 259; Foss, *Nuclear safety and international governance*, 21; Senn, *Gorbachev's failure*, 5.
40. See, for example, Adriana Petryna, *Life exposed: Biological citizens after Chernobyl* (New Jersey: Princeton University Press, 2002), 2; Brown, "In the house that plutonium built," 2.
41. Josephson, *Red atom*, 36; to meet production targets was generally the most important thing in Soviet nuclear industry, not, for example, quality and safety. See Foss, *Nuclear safety and international governance*, 7; see also Herttrich et al., "International co-operation to promote nuclear-reactor safety," 92.
42. Foss, *Nuclear safety and international governance*, 11, note 10.
43. V. Kopustinskas et al., "An approach to estimate radioactive release frequency from Ignalina RBMK-1500 reactor in Lithuania," *Zagadnienia eksploatacji maszyn* 1, no. 149 (2007), 189; Foss, *Nuclear safety and international governance*, 11, 89; Herttrich et al., "International co-operation to promote nuclear-reactor safety," 92.
44. Dawson, *Eco-nationalism*, 3.
45. Urbonavičius, interview.
46. Dawson, *Eco-nationalism*, 41.
47. Foss, *Nuclear safety and international governance*, 2, 15.
48. Wendland, "Inventing the atomograd," 17; Šliavaitė, *From pioneers to target group*, 12; Petryna, *Life exposed*, 4; for the connection between Chernobyl and Ukrainian independence, see Ulrich Beck, *World at risk* (Cambridge: Polity Press, 2009), 171f.; for the connection made in Belarus, see Kasperski, "Chernobyl's aftermath in political symbols."

49. Dawson, *Eco-nationalism*, 43; Senn, *Gorbachev's failure*, 15.
50. Dalia Gineitiene et al., "Public risk perceptions of nuclear power. The case of Sweden and Lithuania," in *Social Processes and the Environment. Lithuania and Sweden*, ed. Anna-Lisa Lindén and Leonardas Rinkevičius (Lund: Lund University, 1999), 149.
51. Alfred Erich Senn, *Lithuania awakening* (Berkeley: University of California Press, 1990), 64; Leonardas Rinkevičius, "Attitudes and values of the Lithuanian green movement in the period of transition," *Filosofija, sociologija*, no. 1 (2001), 75.
52. Zigmas Vaišvila, interview, Vilnius, November 24, 2010.
53. Dawson, *Eco-nationalism*, 36f.
54. Ibid., 34ff.
55. Ibid., 41f.; Senn, *Gorbachev's failure*, 9.
56. Rinkevičius, "Attitudes and values," 72; see also Senn, *Lithuania awakening*, 202f.
57. Dawson, *Eco-nationalism*, 28, 43ff.
58. Ibid., 49.
59. Ibid., 54f.
60. Senn, *Lithuania awakening*, 175.
61. Dawson, *Eco-nationalism*, 187, note 33.
62. Ibid., 48; see also Senn, *Lithuania awakening*, 9, 64.
63. Dawson, *Eco-nationalism*, 52, 170.
64. Senn, *Gorbachev's failure*, xii; Dawson, *Eco-nationalism*, 52.
65. Dawson, *Eco-nationalism*, 30; see also Rinkevičius, "Attitudes and values," 73f.; Senn, *Lithuania awakening*, 125.
66. Dawson, *Eco-nationalism*, 35, 57.
67. Senn, *Gorbachev's failure*, 103f.
68. Dawson, *Eco-nationalism*, 3, 29, 166; see also Kasperski, "Chernobyl's aftermath in political symbols," 84.
69. Dawson, *Eco-nationalism*, 23; Senn, *Gorbachev's failure*, xiii.
70. Cited in Šliavaitė, *From pioneers to target group*, 133.
71. P. Butcher et al., "TSO assistance towards the improvement of nuclear safety in Lithuania. Achievements and perspectives," Paper presented at the Eurosafe forum 2001, Paris, France: 2001.
72. Foss, *Nuclear safety and international governance*, 18.
73. Josephson, *Red atom*, 284.
74. Šliavaitė, *From pioneers to target group*, 57; in 2006–2007 only 3 percent of the employees at the Ignalina nuclear power plant spoke Lithuanian sufficiently well. See Labanauskas, "Social aspects of the functioning of the Ignalina nuclear power plant," 80, 83.
75. Šliavaitė, *From pioneers to target group*, 58, 110, 137.
76. Ibid., 28, 63f., 85.
77. The nostalgia was also a question of relating one's own biography to the history of the towns, since many of the inhabitants arrived in their twenties. The new borders furthermore made it more

expensive and difficult to travel between Lithuania and Russia, which implied disconnections between many people in Sniečkus and their relatives still living in the Russian cities from where they once came. See ibid., 28f., 100, 134; see also Josephson, *Red atom*, 38; Foss, *Nuclear safety and international governance*, 22; Joy Parr, *Sensing changes: Technologies, environments, and the everyday, 1953–2003* (Vancouver: UBC Press, 2010), 8; Brown, "Gridded lives," 41.
78. Foss, *Nuclear safety and international governance*, 21, 53, 58f., 62.
79. Ševaldin, interview; Josephson, *Red atom*, 36.
80. Ševaldin, interview.
81. Butcher et al., "TSO assistance towards the improvement of nuclear safety"; Foss, *Nuclear safety and international governance*, 92.
82. Högselius, "Connecting East and West?" 245, 270; Butcher et al., "TSO assistance towards the improvement of nuclear safety."
83. For example, one of the leading figures in Žemyna became prime minister and another one became personal advisor to the president. See Dawson, *Eco-nationalism*, 31, 59f.
84. Ibid.; Foss, *Nuclear safety and international governance*, 75; Högselius, "Connecting East and West?" 245.
85. Högselius, "Connecting East and West?" 252ff., 264ff.
86. Herttrich et al., "International co-operation to promote nuclear-reactor safety," 94; Foss, *Nuclear safety and international governance*, 25.
87. Cited in Foss, *Nuclear safety and international governance*, 53.
88. Ibid., 25ff.
89. Quote from Butcher et al., "TSO assistance towards the improvement of nuclear safety"; Renata Karaliūtė, "Nuclear knowledge management and preservation in Lithuania," *International Journal of Nuclear Knowledge Management* 1, no. 3 (2005), 218.
90. Dawson, *Eco-nationalism*, 61; Gineitiene et al., "Public risk perceptions," 160.
91. Foss, *Nuclear safety and international governance*, 40; Ševaldin, interview; Gunnar Johanson, interview.
92. Butcher et al., "TSO assistance towards the improvement of nuclear safety."
93. Kopustinskas et al., "An approach to estimate radioactive release," 189.
94. Foss, *Nuclear safety and international governance*, 38ff.
95. Ševaldin, interview; Jan Nistad, interview, Stockholm, October 19, 2010; Gunnar Johanson, interview.
96. Leif Öst, interview, Barsebäck, November 3, 2010.
97. Gunnar Johanson, interview.
98. Ševaldin, interview; Šliavaitė, *From pioneers to target group*, 93.
99. Foss, *Nuclear safety and international governance*.

100. Butcher et al., "TSO assistance towards the improvement of nuclear safety."
101. Karaliūtė, "Nuclear knowledge management"; Butcher et al., "TSO assistance towards the improvement of nuclear safety"; Dawson, *Eco-nationalism*, 61.
102. Nistad, interview.
103. Butcher et al., "TSO assistance towards the improvement of nuclear safety."
104. Foss, *Nuclear safety and international governance*, 3.
105. Ibid., 84.
106. Gunnar Johanson, interview.
107. In 2004, the European Union enlarged by adding five new Central and Eastern European members operating nuclear power stations: Czech Republic, Hungary, Lithuania, Slovakia, and Slovenia. All these nations relied on nuclear power for at least a quarter of their electricity needs. See Jane I. Dawson and Robert G. Darst, "Meeting the challenge of permanent nuclear waste disposal in an expanding Europe: Transparency, trust and democracy," *Environmental Politics* 15, no. 4 (2006), 610, 625.
108. Jonas Gylys and Leonas Ašmantas, "The specific nuclear energy problems in Lithuania," Paper presented at the 19th World Energy Congress, Sydney, Australia, 2004; Karaliūtė, "Nuclear knowledge management," 221.
109. Gylys and Ašmantas, "The specific nuclear energy problems," 5.
110. Wendland, "Inventing the atomograd," 17.
111. Šliavaitė, *From pioneers to target group*, 44, 88, 142, 176f.
112. Ibid., 149.
113. Ibid., 87, 91, 102, 151; Urbonavičius, interview.
114. Joanna Bourke, *Fear: A cultural history* (London: Virago, 2005); Beck, *World at risk*; Spencer R. Weart, *The rise of nuclear fear* (Cambridge, MA: Harvard University Press, 2012); Charles Perrow, *Normal accidents: Living with high-risk technologies* (Princeton, NJ: Princeton University Press, 1999); Paul Slovic, *The perception of risk* (London: Earthscan Publications, 2000); Slovic cited in Dawson and Darst, "Meeting the challenge of permanent nuclear waste disposal," 23.
115. Gineitiene et al., "Public risk perceptions," 154.
116. Ibid., 153.
117. Ibid., 124.
118. Šliavaitė, *From pioneers to target group*, 87, 92; Dawson, *Eco-nationalism*, 39.
119. Šliavaitė, *From pioneers to target group*, 94.
120. Ibid., 47, 96.
121. Wendland, "Inventing the atomograd," 17; Šliavaitė, *From pioneers to target group*, 12.

122. Schmid, "Celebrating tomorrow today," 342ff.
123. Brown, "In the house that plutonium built."
124. Šliavaitė, *From pioneers to target group*, 46; Senn, *Lithuania awakening*, 9f., 202f.
125. Josephson, *Red atom*, 32, 38; Wendland, "Inventing the atomograd."
126. Brown, "In the house that plutonium built," 3, 26, 36.
127. Ibid., 25.
128. Wendland, "Inventing the atomograd," 18; Stefan Buzar, *Energy poverty in Eastern Europe: Hidden geographies of deprivation* (Aldershot: Ashgate, 2007).
129. Josephson, *Red atom*, 290; Karaliūtė, "Nuclear knowledge management," 222.
130. Marija Drėmaitė, "Industrial heritage in a rural country: Interpreting the industrial past in Lithuania," in *Industrial heritage around the Baltic Sea*, ed. Marie Nisser et al., Uppsala Studies in Economic History 92 (Uppsala: Acta Universitatis Upsaliensis, 2012), 65f.
131. Cinis et al., "Perfect representations of Soviet planned space," 227.
132. Drėmaitė, "Industrial heritage in a rural country," quote from 73.
133. Within this trend, some nuclear power plants have been reused for cultural and commercial uses, although this is rather rare. One example is a breeder reactor site that was never in operation, Kalkar in Germany, now transformed into an amusement park (including a museum of the site's history). See Maja Fjæstad, "Ett kärnkraftverk återuppstår: Från Snr300 till Wunderland Kalkar," *Bebyggelsehistorisk tidskrift*, no. 63 (2012).
134. Drėmaitė, "Industrial heritage in a rural country," 65f.
135. Algimantas Degutis (with Alfredas Jomantas), interview, Vilnius, November 26, 2010; Ševaldin, interview.
136. Ševaldin, interview.
137. Drėmaitė, "Industrial heritage in a rural country," quote from 73.
138. Schmid, "Celebrating tomorrow today," 335ff.
139. Degutis, interview.
140. Ibid.
141. Ibid.
142. Ševaldin, interview.
143. Urbonavičius, interview.
144. Ibid.
145. Kenneth E. Foote, *Shadowed ground: America's landscapes of violence and tragedy*, revised edition (Austin: University of Texas Press, 2003), 29.
146. The construction of reactor 3 started in 1985, was suspended in 1988, and demolition began in 1989. Dismantling was completed in 2008; for a discussion on decommissioning costs, see Åsa Moberg, *Ett extremt dyrt och livsfarligt sätt att värma vatten: En bok om kärnkraft* (Stockholm: Natur och kultur, 2014), 246.

147. Urbonavičius, interview.
148. Ševaldin, interview.
149. Urbonavičius, interview.
150. David E. Nye and Sarah Elkind, eds., *The anti-landscape* (Amsterdam: Rodopi, 2014).
151. See also Gunilla Bandolin and Sverker Sörlin, *Laddade landskap – värdering och gestaltning av teknologiskt sublima platser*, R-07–14 (Stockholm: SKB, 2007).
152. Paul Ricœur, *Memory, history, forgetting*. Translated by Kathleen Blamey and David Pellauer (Chicago: University of Chicago Press, 2004).

5 Industrial Nature

1. "Industrial nature" can also be understood as "industrialized nature," that is, environments directly changed by industrial production. See, for example, Paul R. Josephson, *Industrialized nature: Brute force technology and the transformation of the natural world* (Washington DC: Island Press, 2002); Gene Desfor and Lucian Vesalon, "Urban expansion and industrial nature: A political ecology of Toronto's port industrial district," *International Journal of Urban and Regional Research* 32, no. 3 (2008); however, for the sake of this chapter, I choose to leave this understanding aside and concentrate on conceptualizations emphasizing a post-industrial situation or industrial wastelands, beside production or after production has ceased.
2. Tobias Plieninger and Claudia Bieling, "Resilience and cultural landscapes: Opportunities, relevance and ways ahead," in *Resilience and the cultural landscape: Understanding and managing change in human-shaped environments*, ed. Tobias Plieninger and Claudia Bieling (Cambridge: Cambridge University Press, 2012), 332; Thymio Papayannis and Peter Howard, "Editorial: Nature as heritage," *International Journal of Heritage Studies* 13, no. 4–5 (2007), 300; Hugh Cheape, Mary-Cate Garden, and Fiona McLean, "Editorial: Heritage and the environment," *International Journal of Heritage Studies* 15, no. 2–3 (2009); Andrew Karvonen, *Politics of urban runoff: Nature, technology and the sustainable city* (Cambridge, MA: MIT Press, 2011), quote from 188.
3. Anne Brownley Raines, "Wandel durch (Industrie) Kultur [Change through (industrial) culture]: Conservation and renewal in the Ruhrgebiet," *Planning Perspectives* 26, no. 2 (2011), 195f.
4. Nate Millington, "Post-industrial imaginaries: Nature, representation and ruin in Detroit, Michigan," *International Journal of Urban and Regional Research* 37, no. 1 (2013), 285, 91.
5. Gaëlle Covo, "Spatial planning, structural change and regional development policies within the Ruhr area in Germany," in *Exploring the Ruhr in Germany* (Bochum, 2001); Andreas Keil and Burkhard Wetterau, *Metropolis Ruhr: A regional study of the new Ruhr* (Essen: Regionalverband Ruhr, 2013), 14f.

6. Joachim Huske, *Die Steinkohlenzechen im Ruhrrevier. Daten und Fakten von den Anfängen bis 2005, 3. überarbeitete und erweiterte Auflage* (Bochum: Deutsches Bergbau-Museum, 2006; repr., 3rd).
7. Ingerid Helsing Almaas, "Regenerating the Ruhr—IBA Emscher Park project for the regeneration of Germany's Ruhr region," *The Architectural Review* CCV, no. 1224 (1999).
8. Jörg Dettmar, "Forests for shrinking cities? The project 'Industrial Forests of the Ruhr,'" in *Wild urban woodlands: New perspectives for urban forestry*, ed. Ingo Kowarik and Stefan Körner (Berlin, Heidelberg: Springer-Verlag Berlin Heidelberg, 2005), 264; see also Keil and Wetterau, *Metropolis Ruhr*, 26.
9. Polise Moreira De Marchi, "Ruhrgebiet: redesigning an industrial region," in *Exploring the Ruhr in Germany* (Bochum, 2001), 29.
10. Wolfram Höfer and Vera Vicenzotti, "Post-industrial landscapes: Evolving concepts," in *The Routledge companion to landscape studies*, ed. Peter Howard, Ian Thompson, and Emma Waterton (Abingdon: Routledge, 2013), 406f.; Ingo Kowarik and Andreas Langer, "Natur-Park Südgelände: Linking conservation and recreation in an abandoned railyard in Berlin," in *Wild urban woodlands: New perspectives for urban forestry*, ed. Ingo Kowarik and Stefan Körner (Berlin, Heidelberg: Springer-Verlag Berlin Heidelberg, 2005), 287; Jens Lachmund, *Greening Berlin: The co-production of science, politics, and urban nature* (Cambridge, MA: MIT Press, 2013).
11. Kowarik and Langer, "Natur-Park Südgelände," 287.
12. Ibid.
13. Ingo Kowarik, "Novel urban ecosystems, biodiversity, and conservation," *Environmental Pollution*, no. 159 (2011), 6.
14. Ingo Kowarik, "Wild urban woodlands: Towards a conceptual framework," in *Wild urban woodlands: New perspectives for urban forestry*, ed. Ingo Kowarik and Stefan Körner (Berlin, Heidelberg: Springer-Verlag Berlin Heidelberg, 2005), 2.
15. Heiderose Kilper and Gerald Wood, "Restructuring policies: The Emscher park international building exhibition," in *The rise of the rustbelt*, ed. Philip Cooke (London: UCL Press, 1995), 226; Wolfgang Pehnt, "Changes have to take place in people's heads first," *Topos: European landscape magazine*, no. 26 (1999), 22. Kersti Morger's private collections.
16. *Internationale Bauausstellung Emscher-Park: Werkstatt für die Zukunft alter Industriegebiete*, Memorandum zu Inhalt und Organisation (Düsseldorf: Der Minister für Stadtentwicklung, Wohnen und Verkehr des Landes Nordrhein-Westfalen, 1988). Kersti Morger's private collections.
17. Covo, "Spatial planning, structural change and regional development policies."
18. Thomas Sieverts, "Die Internationale Bauausstellung Emscher Park: Werkstatt zur Erneuerung alter Industriegebiete—Eine

strukturpolitische Initiative des Landes Nordrhein-Westfalen," *Arcus Architektur und Wissenschaft: IBA Emscher Park, Zukunkftswerkstatt für Industrieregionen*, no. 13 (1991), 6. Kersti Morger's private collections.
19. Robert Shaw, "The International Building Exhibition (IBA) Emscher Park, Germany: A model for sustainable restructuring?," *European Planning Studies* 10, no. 1 (2002), 82.
20. Ibid., 84; Gerhard Seltmann, "Eine unmögliche Ausstellung? Merkzeichen für die Internationale Bauausstellung Emscher Park," *Arcus Architektur und Wissenschaft: IBA Emscher Park, Zukunkftswerkstatt für Industrieregionen*, no. 13 (1991). Kersti Morger's private collections.
21. Raines, "Wandel durch (Industrie) Kultur," 200.
22. Joachim Weiss et al., "Nature returns to abandoned industrial land: Monitoring succession in urban-industrial woodlands in the German Ruhr," in *Wild urban woodlands: New perspectives for urban forestry*, ed. Ingo Kowarik and Stefan Körner (Berlin, Heidelberg: Springer-Verlag Berlin Heidelberg, 2005), 143.
23. Wolfgang Ebert, interview, Geldern, February 12, 2007; see also Rainer Slotta, "Industrial archaeology in the Federal Republic of Germany," in *Industrial heritage '84 national reports: The fifth international conference on the conservation of the industrial heritage* (Washington DC: 1984), 40.
24. Raines, "Wandel durch (Industrie) Kultur," 195; Ebert, interview.
25. "A selection of projects: IBA Emscher Park," *Topos: European landscape magazine*, no. 26 (1999), 98. Kersti Morger's private collections; Local governments and private companies often jointly financed the projects within IBA Emscher Park. However, at Landschaftspark Duisburg-Nord most of the expenditures were publicly funded. See Environmental Protection Agency, International brownfields case study, Emscher Park, Germany, www.epa.gov (accessed October 15, 2007).
26. Ebert, interview; the German Society for Industrial History was founded in 1986.
27. Ibid.
28. Environmental Protection Agency, International Brownfields Case Study, Westergasfabriek, Amsterdam, Netherlands. www.epa.gov (accessed October 15, 2007); Andreas Keil, "Use and perception of post-industrial urban landscapes in the Ruhr," in *Wild urban woodlands: New perspectives for urban forestry*, ed. Ingo Kowarik and Stefan Körner (Berlin, Heidelberg: Springer-Verlag Berlin Heidelberg, 2005), 128; Anna Storm, *Koppardalen: Om historiens plats i omvandlingen av ett industriområde*, Stockholm papers in the history and philosophy of technology, Trita-HOT: 2047 (Stockholm: KTH lic., 2005).
29. Ebert, interview; nevertheless, the word "park" could also signify ongoing industrial activity. A steel works in northern Duisburg was

for example called "Industriepark Ost." See Håkon W. Andersen et al., *Fabrikken* (Oslo: Scandinavian Academic Press/Spartacus Forlag, 2004), 576.
30. The CSR triangle theory is an ecological classification of plants into competitors, stress-tolerators and ruderals. Ruderals are defined as having "high reproductive rates in transient disturbed habitats." See J. Philip Grime and Simon Pierce, *The evolutionary strategies that shape ecosystems* (John Wiley & Sons, Ltd, 2012), 12f.
31. Personal communication with Arne Anderberg, September 24, 2013; Encyclopedia Britannica http://www.britannica.com/ (accessed September 25, 2013); Oxford English Dictionary online, entry for "ruderal," http://www.oed.com/view/Entry/276667?redirectedFrom=ruderal& (accessed September 25, 2013).
32. Peter Latz, "Landscape Park Duisburg-Nord: The metamorphosis of an industrial site," in *Manufactured sites: Rethinking the post-industrial landscape*, ed. Niall Kirkwood (New York: Spon Press, 2001), 157.
33. Latz und Partner, www.latzundpartner.de/projects/detail/18 (accessed September 25, 2013).
34. Landschaftspark Duisburg-Nord, www.landschaftspark.de (accessed March 19, 2007).
35. Personal communication with Dietrich Soyez, January 22, 2014.
36. For an analysis of different experiences of nature including examples from the Ruhr area, see Klas Sandell, "Var ligger utomhus? Om nya arenor för natur- och landskapsrelationer," in *Utomhusdidaktik*, ed. Iann Lundegård, Per-Olof Wickman, and Ammi Wohlin (Lund: Studentlitteratur, 2004).
37. Dettmar, "Forests for shrinking cities?" 265.
38. Ibid.; quote from Ebert, interview.
39. Ebert, interview.
40. John Urry, "The place of emotions within place," in *Emotional geographies*, ed. Joyce Davidson, Liz Bondi, and Mick Smith (Aldershot: Ashgate, 2005), quote from 81.
41. Shaw, "The International Building Exhibition (IBA) Emscher Park, Germany," 91; see also Axel Föhl, "'Ein Abwurfplatz für Meteoriten'. Industriedenkmale und die IBA," *Industrie-Kultur* 3 (1999); Ulrich Heinemann, "Industriekultur: Vom Nutzen zum Nachteil für das Ruhrgebiet," *Forum Industriedenkmalpflege und Geschichtskultur* 1 (2003).
42. Ursula Poblotzki, "Transformation of a landscape," *Topos: European landscape magazine*, no. 26 (1999), 48. Kersti Morger's private collections.
43. Keil, "Use and perception of post-industrial urban landscapes in the Ruhr."
44. Ibid., 128.

NOTES

45. Ibid., 119. Their field studies were carried out in 1997, 1998, and 2003.
46. Ibid., 119, 27.
47. Ibid., 127.
48. Ibid., 125.
49. "A selection of projects: IBA Emscher Park," 126. Kersti Morger's private collections.
50. Ebert, interview; the figure is specified in Marie Nisser, "Nytt liv i Europas gamla industriområden," *Kulturmiljövård*, no. 6 (1994), 15.
51. Raines, "Wandel durch (Industrie) Kultur," 189f.
52. Ibid., 191.
53. Ibid., 186ff.
54. Dagmar Kift, "Heritage and history: Germany's industrial museums and the (re-) presentation of labour," *International Journal of Heritage Studies* 17, no. 4 (2011), 380; Raines, "Wandel durch (Industrie) Kultur," 186ff.; Rita Müller, "Museums designing for the future: Some perspectives confronting German technical and industrial museums in the twenty-first century," *International Journal of Heritage Studies* 19, no. 5 (2013), 511.
55. Kift, "Heritage and history," 385f.
56. Ibid., 380; for a slightly different perspective, see Müller, "Museums designing for the future," 511.
57. Shiloh R. Krupar, *Hot spotter's report: Military fables of toxic waste* (Minneapolis: University of Minnesota Press, 2013).
58. Ibid., 6–11.
59. Millington, "Post-industrial imaginaries," 283–89.
60. Ibid., 293.
61. See, for example, Chris Hagerman, "Shaping neighborhoods and nature: Urban political ecologies of urban waterfront transformations in Portland, Oregon," *Cities* 24, no. 4 (2007), 285ff., quote from 290; Anna Storm, *Hope and rust: Reinterpreting the industrial place in the late 20th century* (Stockholm: KTH diss., 2008).
62. Hagerman, "Shaping neighborhoods and nature," 288.
63. Ibid., 287–95, quote from 295.
64. Storm, *Hope and rust*; Storm, *Koppardalen*.
65. Höfer and Vicenzotti, "Post-industrial landscapes," 405.
66. Ibid., 406.
67. Ibid.
68. Thomas Lekan and Thomas Zeller, *Germany's nature: Cultural landscapes and environmental history* (New Brunswick, NJ: Rutgers University Press, 2005), 4; a similar logic underpins the volume *Urban wildscapes*, among other things including examples from several European countries. Anna Jorgensen and Richard Keenan, ed., *Urban wildscapes* (New York: Routledge, 2011).

69. Höfer and Vicenzotti, "Post-industrial landscapes," 409.
70. Ibid., 410.
71. Ebert, interview; Dietrich Soyez, "Das amerikanische Industriemuseum 'Sloss Furnaces'—ein Modell für das Saarland?," *Annales. Forschungsmagazin der Universität des Saarlandes*, no. 1 (1988).
72. Hagerman, "Shaping neighborhoods and nature," 295.
73. Divya P. Tolia-Kelly, "Landscape and memory," in *The Routledge companion to landscape studies*, ed. Peter Howard, Ian Thompson, and Emma Waterton (Abingdon: Routledge, 2013), 323.
74. Zygmunt Bauman, "From pilgrim to tourist—or a short history of identity," in *Questions of cultural identity*, ed. Stuart Hall and Paul du Gay (London: SAGE Publications Ltd, 1996), 30; Sharon Zukin, *Loft living: Culture and capital in urban change* (New Brunswick, NJ: Rutgers University Press, 1982), 111f.; the conversion of former industrial buildings for cultural purposes is taking place in many places in Europe. See, for example, Fazette Bordage, ed., *The factories: Conversions for urban culture* (Basel: Birkhäuser, 2002).
75. Millington, "Post-industrial imaginaries," 281; Tim Edensor, *Industrial ruins: Spaces, aesthetics, and materiality* (Oxford: Berg, 2005).
76. Tim Edensor, *Industrial ruins*; see also Richard Pettersson, "Plats, ting och tid: Om autenticitetsuppfattning utifrån exemplet Rom," in *Topos: Essäer om tänkvärda platser och platsbundna tankar* (Stockholm: Carlssons Bokförlag, 2006); Klas Sandell, "Friluftsliv, platsperspektiv och samhällskritik," in *Topos: Essäer om tänkvärda platser och platsbundna tankar* (Stockholm: Carlssons Bokförlag, 2006).
77. *Internationale Bauausstellung Emscher-Park*, 44. Kersti Morger's private collections.
78. Jan Turtinen, *Världsarvets villkor: Intressen, förhandlingar och bruk i internationell politik*, Acta Universitatis Stockholmiensis, Stockholm studies in Ethnology 1 (Stockholm: Stockholm University, 2006), 60f.; the concept of authenticity is under negotiation, however, especially under the influence of value systems in Asia. See, for example, the Nara Document on Authenticity, www.iicc.org.cn/Info.aspx?ModelId=1&Id=422 (accessed April 12, 2014).
79. Kowarik, "Wild urban woodlands."
80. Ibid., 17ff.
81. Ibid., 21.
82. Ibid., 22; see also Jorgensen and Keenan, eds., *Urban wildscapes*.
83. Kowarik and Langer, "Natur-Park Südgelände," 288, 90.
84. Ibid., 291.
85. Dettmar, "Forests for shrinking cities?," 267.
86. Ibid.
87. For other large-scale transformations in the Ruhr, see, for example, Keil and Wetterau, *Metropolis Ruhr*, 30.

88. Raines, "Wandel durch (Industrie) Kultur," 197f.; Müller, "Museums designing for the future," 525; see also Völklinger Hütte, www .voelklinger-huette.org/de/faszination-weltkulturerbe/das-paradies/ and Das Paradies, www.euromuse.net/en/exhibitions/exhibition /view-e/das-paradies/ (both sites accessed August 29, 2013).
89. Kowarik and Langer, "Natur-Park Südgelände," 296; Dettmar, "Forests for shrinking cities?" 269.
90. One could note how Germany became industrialized and urbanized later and more rapidly than many other countries like France and Britain, and also that it experienced six different political regimes from 1871 up to 1990. See Lekan and Zeller, *Germany's nature*, 4.
91. Gregory Ashworth and Peter Howard, *European heritage planning and management* (Exeter: Intellect Ltd, 1999), 24.
92. David Lowenthal, *The heritage crusade and the spoils of history* (Cambridge: Cambridge University Press, 1998), x.
93. Zukin, *Loft living*, 73.
94. Kenneth E. Foote, *Shadowed ground: America's landscapes of violence and tragedy*, revised edition (Austin: University of Texas Press, 2003), 25.
95. Kowarik, "Wild urban woodlands," 3.
96. Cited in Raines, "Wandel durch (Industrie) Kultur," 203; see also Karl Ganser, "The present state of affairs...Karl Ganser in a conversation with Carl Steckeweh and Kunibert Wachten," in *Change without growth? Sustainable urban development for the 20st century. VI. Architecture biennale Venice 1996*, edited by Kunibert Wachten (Braunschweig/Wiesbaden: Friedr. Vieweg & Sohn Verlagsgesellschaft mbH, 1996), 12–23.
97. Shaw, "The International Building Exhibition (IBA) Emscher Park, Germany," 84f.
98. Huske, *Die Steinkohlenzechen im Ruhrrevier. Daten und Fakten von den Anfängen bis 2005, 3. überarbeitete und erweiterte Auflage*.
99. Raines, "Wandel durch (Industrie) Kultur," 193, quote from 201.
100. Ibid., 183.
101. Ibid., 195.
102. Dettmar, "Forests for shrinking cities?," 272.
103. Landschaftspark Duisburg-Nord, www.landschaftspark.de/architektur-natur and www.landschaftspark.de/architektur-natur/landschaftsarchitektur (accessed September 25, 2013).
104. Quote from Universität Oldenburg, landscape ecology, www .landeco.uni-oldenburg.de/20716.html (accessed September 25, 2013); Mira Kattwinkel, Robert Biedermann, and Michael Kleyer, "Temporary conservation for urban biodiversity," *Biological Conservation*, no. 144 (2011).
105. David Harvey, "Emerging landscapes of heritage," in *The Routledge companion to landscape studies*, ed. Peter Howard, Ian Thompson, and Emma Waterton (Abingdon: Routledge, 2013), 154f.

106. Matthew Gandy, "Urban nature and the ecological imaginary," in *In the nature of cities: Urban political ecology and the politics of urban metabolism*, ed. Nikolas C. Heynen, Maria Kaika, and Erik Swyngedouw (New York, NY: Routledge, 2006), 71.
107. Mattias Qviström, "Network ruins and green structure development: An attempt to trace relational spaces of a railway ruin," *Landscape Research* 37, no. 3 (2011), 257.
108. Millington, "Post-industrial imaginaries," 282, 86.
109. Karvonen, *Politics of urban runoff*, 188.

6 Enduring Spirit

1. Eva Vikström, *Industrimiljöer på landsbygden: Översikt över kunskapsläget* (Stockholm: Riksantikvarieämbetet, 1994), 42–73; Eva Vikström, *Platsen, bruket och samhället: Tätortsbildning och arkitektur 1860–1970* (Stockholm: Statens råd för byggnadsforskning, 1991), 23ff.
2. Ewa Bergdahl, Maths Isacson, and Barbro Mellander, eds., *Bruksandan—hinder eller möjlighet? Rapport från en seminarieserie i Bergslagen*, Ekomuseum Bergslagens skriftserie no 1 (Smedjebacken: Ekomuseum Bergslagen, 1997).
3. Maths Isacson, *Industrisamhället Sverige: Arbete, ideal och kulturarv* (Lund: Studentlitteratur, 2007), 195, 275; Maths Isacson, "Bruksandan—hinder eller möjlighet? Sammanfattning av fyra seminarier i Bergslagen 1995–1997," in *Bruksandan—hinder eller möjlighet? Rapport från en seminarieserie i Bergslagen. Ekomuseum Bergslagens skriftserie no 1*, ed. Ewa Bergdahl, Maths Isacson, and Barbro Mellander (Smedjebacken: Ekomuseum Bergslagen, 1997), 127.
4. Einar Lövgren, *Folkare din hembygd* (Avesta, 1977), 120.
5. Helena Kåks, *Avesta: Industriarbete och vardagsliv genom 400 år*, DFR-rapport 2002:1 (Falun: Dalarnas forskningsråd, 2002), 27, 37.
6. Sten von Matérn, "Hållpunkter i Avestas historia," in *Koppardalens utveckling: Underlag för prioritering och detaljerad planering av Koppardalens vidare utveckling och exploatering*, 1996, 2. Avesta municipality.
7. Eva Vikström, *Bruksandan och modernismen: Brukssamhälle och folkhemsbygge i Bergslagen 1935–1975*, Nordiska museets handlingar 126 (Stockholm, 1998), 171–78.
8. *Detaljplan för Koppardalens industriområde, planbestämmelser* (July 26, 1989). Avesta municipality.
9. Vikström, *Bruksandan och modernismen*, 174ff.
10. Eva Rudberg, *Alvar Aalto i Sverige*, Arkitekturmuseets skriftserie nr 12 (Stockholm: Arkitekturmuseet, 2005), 66–77.

11. Isacson, *Industrisamhället Sverige*, 219; Jan Fredriksson, "Alvar Aalto och den uteblivna förnyelsen av Avesta stadscentrum," *Dalarna* (1997).
12. *Svenska Dagbladet* (November 3, 1976). Tidningsurklipp 1976–1983, F1:10, Handlingar angående Avesta, Bo Hermelins samling. Nordstjernan Corporation central archives, Engelsbergsarkivet; "Anteckningar rörande Avesta Jernverk under disponent Walfrid Erikssons tid (1927–49), Meddelande 18.2." 1971. Museifrågor, F2B:119, Avesta Jernverks AB. Nordstjernan Corporation central archives, Engelsbergsarkivet; Cyber City web encyclopedia, http://urbanhistory.historia.su.se/cybercity/stad/avesta/befolkning.htm (accessed November 12, 2013), category "tätort" meaning "densely built-up area."
13. "Avtal mellan Avesta AB och Avesta Industristad AB. Bilaga 2, Hyreskontrakt 1." (1986) 3. Avesta municipality.
14. Marie Nisser, "Industriminnen under hundra år," *Nordisk Museologi*, no. 1 (1996).
15. "Utredning beträffande erforderligt utrymme för museisamlingarna 1944: Förslag till allmän plan för bergstekniskt och brukshistoriskt museum i Avesta. Meddelande från Bo Hermelin, LAP 43/69" (October 3, 1969). Handlingar angående Avesta, F1:6. Nordstjernan Corporation central archives, Engelsbergsarkivet; see also "Några industrihistoriska minnesmärken" (1946). Handlingar i diverse ämnen, F5B:6, Kultur- och industrihistoriska minnesmärken. Nordstjernan Corporation central archives, Engelsbergsarkivet.
16. Isacson, *Industrisamhället Sverige*, 217.
17. Rudberg, *Alvar Aalto i Sverige*.
18. Nisser, "Industriminnen under hundra år;" see also Marie Nisser and Fredric Bedoire, "Industrial monuments in Sweden: Conservation, documentation and research," in *The industrial heritage: The third international conference on the conservation of industrial monuments. Transactions 2. Scandinavian reports*, ed. Marie Nisser (Stockholm: Nordiska museet, 1978), 74; Anders Houltz, *Teknikens tempel: Modernitet och industriarv på Göteborgsutställningen 1923*, Stockholm papers in the history and philosophy of technology, Trita-HOT: 2041 (Hedemora: Gidlund, 2003).
19. Annika Alzén, *Fabriken som kulturarv: Frågan om industrilandskapets bevarande i Norrköping 1950–1985* (Stockholm/Stehag: Brutus Östlings Bokförlag Symposion, 1996), 23f.; Nisser and Bedoire, "Industrial monuments in Sweden," 74ff.
20. Maths Isacson, "Industrisamhällets faser och industriminnesforskningens uppgifter," in *Industrins avtryck: Perspektiv på ett forskningsfält*, ed. Dag Avango and Brita Lundström (Stockholm/Stehag: Symposion, 2003), quote from 21; Nisser, "Industriminnen under hundra år," 79; Marie Nisser, "Industriminnen på den

internationella arenan," in *Industrisamhällets kulturarv: Betänkande av Delegationen för industrisamhällets kulturarv* (Stockholm: Statens offentliga utredningar (SOU) 2002: 67, 2002), 220.
21. "Skrota eller underhålla masugnspipor och ruiner?," *Falu Kuriren* (September 3, 1962); Isacson, "Industrisamhällets faser," 21; Nisser, "Industriminnen under hundra år," 79.
22. "PM angående äldre anläggningar inom norra verksområdet i Avesta, docent Marie Nisser" (June 15, 1987). Karin Perers' private collections.
23. "Masugnsbyggnaden såsom museum. Meddelande från Bo Hermelin till VD m.fl." (March 4, 1970). Avesta Jernverks AB. Nordstjernan Corporation central archives, Engelsbergsarkivet; see also "Betr. museum i hyttan. Meddelande från Bo Hermelin till VD och EZ" (May 20, 1970). Avesta Jernverks AB. Nordstjernan Corporation central archives, Engelsbergsarkivet.
24. Jan Burell, interview, Avesta, April 26, 2006; Lars Åke Everbrand, interview, Avesta, April 16, 2003 and May 18, 2006.
25. The reasons for this are still unknown to me.
26. Anna Storm, *Koppardalen: Om historiens plats i omvandlingen av ett industriområde*, Stockholm papers in the history and philosophy of technology (Stockholm: KTH lic., 2005).
27. Barbro Bursell, "Bruket och bruksandan," in *Bruksandan— hinder eller möjlighet? Rapport från en seminarieserie i Bergslagen. Ekomuseum Bergslagens skriftserie no 1*, ed. Ewa Bergdahl, Maths Isacson, and Barbro Mellander (Smedjebacken: Ekomuseum Bergslagen, 1997), 10.
28. Therese Nordlund, *Att leda storföretag: En studie av social kompetens och entreprenörskap i näringslivet med fokus på Axel Ax:son Johnson och J. Sigfrid Edström, 1900–1950*, Stockholm studies in economic history (Stockholm: Acta Universitatis Stockholmiensis, 2005), 114ff., 258.
29. Ibid., 255f.
30. Ibid., 255.
31. Letter from 1931, quoted in ibid.
32. Ibid., 263.
33. "Avtal mellan Avesta AB och Gamla Byn AB. Köpekontrakt" (1986). Avesta municipality; "Avtal mellan Avesta AB och Avesta Industristad AB. Huvudavtal. Letter of intent" (June 4, 1986). Avesta municipality; "Avtal mellan Avesta AB och Avesta Industristad AB. Ramavtal" (1986), 1. Avesta municipality.
34. "Avtal mellan Avesta AB och Avesta Industristad AB. Bilaga 1, Överenskommelse om fastighetsreglering" (1986), 1f. Avesta municipality; "Avtal mellan Avesta AB och Avesta Industristad AB. Bilaga 1:14, Asbestförekomst i ventilationsanläggningar—Avesta AB i Avesta" (1985). Avesta municipality; "Avtal mellan Avesta AB och Avesta Industristad AB. Bilaga 1:15, Angående asbest i Norra Verken" (1986). Avesta municipality.

35. Kåks, *Avesta*, 137; Cyber City web encyclopedia, http://urbanhistory.historia.su.se/cybercity/stad/avesta/befolkning.htm (accessed November 12, 2013), category "tätort" meaning "densely built-up area."
36. Per-Erik Pettersson, interview, Avesta, May 17, 2006; Kenneth Linder, interview, Avesta, May 18, 2006; Åke Johansson, interview, Avesta, May 18, 2006.
37. Kåks, *Avesta*, 137ff; see also Gunnel Forsberg, "Reproduktionen av den patriarkala bruksandan," in *Bruksandan—hinder eller möjlighet? Rapport från en seminarieserie i Bergslagen. Ekomuseum Bergslagens skriftserie no 1*, ed. Ewa Bergdahl, Maths Isacson, and Barbro Mellander (Smedjebacken: Ekomuseum Bergslagen, 1997).
38. Jan Thamsten, interview, Avesta, September 16, 2003.
39. Magnus Eklund, *Att möta krisen: Strukturomvandlingen i Avesta 1960–90 och kommunalt motagerande 1978–86*, C-uppsats (Uppsala universitet, 2000), 25f.; "Koppardalen blir namnet," *Avesta Tidning* (May 19, 1987).
40. Everbrand, interview.
41. "I juni smäller det!," *Dala-Demokraten* (April 16, 1993).
42. Karin Perers, "Avesta Art," *Folkarebygden* (1995), 17f.
43. Ibid., 15.
44. "Hyttan blev konsthall," *Attraktiva Avesta* (2002). Avesta municipality; see also "Avesta Art invigs i morgon," *Avesta Tidning* (May 3, 2002).
45. Perers, "Avesta Art," 15f.
46. Everbrand, interview.
47. Ibid.
48. "Program för utveckling av Koppardalen 1998–2003" (1998). Avesta municipality; "Koppardalens förnyelse: Slutredovisning etapp 1" (2001), 2. Avesta municipality.
49. Isacson, *Industrisamhället Sverige*, 263ff.
50. "Rundvandring i Avesta. Kompendium sammanställt av Avesta kommun." (1993), 27. Avesta municipal archives.
51. "Program för utveckling av Koppardalen 1998–2003," 14. Avesta municipality.
52. "Brev från Ulf Löfwall till Hans Ångman" (September 30, 1994). Dalarna County Administrative Board; "Anhållan om rivningslov från Lars Markström och Hans Ångman, Avesta Industristad AB, till Miljö- och stadsbyggnadskontoret, Avesta kommun" (September 26, 1994). Dalarna County Administrative Board.
53. "Program för utveckling av Koppardalen 1998–2003," 14. Avesta municipality.
54. Caroline Tholander, *Koppardalsprojektet ur ett uthållighetsperspektiv: Begrepps- och verktygslåda. Teorier och erfarenheter från Chalmers, m. fl.*, Uppsats inom kursen "Stadsbyggnad med miljöhänsyn" (Göteborg: CTH, 1999); "Brev från Caroline

Tholander till Riksantikvarieämbetet, nr 331–5213–1999" (October 14, 1999). ATA, the Swedish National Heritage Board Archives; "Tjänsteanteckning, nr 5213–1999" (November 19, 1999). ATA, the Swedish National Heritage Board Archives; "Beslut 1999-11-15. Länsstyrelsen Dalarnas län, kulturmiljöenheten, biträdande länsantikvarie Tommy Nyberg, nr 221–9974–99, 311–5885–1999" (November 18, 1999). ATA, the Swedish National Heritage Board Archives; "Länsrätten i Dalarnas läns dom i mål nr 3005–99, överklagande av Länsstyrelsens beslut den 15 november. Kommunstyrelsen, delgivning" (1999). Avesta municipality; "Kallvalsen blir inte kulturminnesmärke," *Avesta Tidning* (November 22, 1999); Björn Björck et al., eds., *Koppardalens förnyelse, etapp 2. Projekt dokumentation: Identifiering av värdebärare. Koppardalen i förändring* (Avesta: Avesta kommun, 2001), 15.
55. Björck et al., *Koppardalens förnyelse, etapp 2*, 41, 62, 66.
56. Everbrand, interview.
57. "Anteckningar från projektgruppen Koppardalens förnyelse möte den 30 juni 1999." Kommunstyrelsen, delgivning. KK94/0261 (September 2, 1999). Avesta municipality; Johan Hiller et al., "Avesta projekt Kopparlänken dialog" (KTH, 1999).
58. Björck et al., *Koppardalens förnyelse, etapp 2*, 36.
59. Anders Dickfors, Dan Ola Norberg, and Marianne Wahlström, "Koppardalen, Avesta: Skyltning, parkering och markdisposition" (2002). Avesta municipality; Kersti Lenngren, "Gestaltningsprogram del 1 och del 2" (2003). Avesta municipality.
60. "Anteckningar från projektgruppen Koppardalens förnyelse möte den 30 juni 1999." Kommunstyrelsen, delgivning. KK94/0261 (September 2, 1999). Avesta municipality.
61. Kjell Lundberg, "Dialog," *Avesta Tidning* (January 24, 2001).
62. Isacson, *Industrisamhället Sverige*, 238.
63. "Koppardalens förnyelse: Etapp 2, 2000–2003" (2001), 2. Avesta municipality.
64. Lars Åke Everbrand and Dan Ola Norberg, *Koppardalsprojektet inför etapp 2* (Avesta, 2000), 31.
65. Roland Bärtilsson, "Lång debatt om Koppardalsprojekt," *Avesta Tidning* (October 27, 2000).
66. See, for example, Environmental Protection Agency, International Brownfields Case Study, Westergasfabriek, Amsterdam, Netherlands, www.epa.gov (accessed October 15, 2007).
67. Everbrand, interview.
68. Ibid.
69. Ibid.
70. "Koppardalens förnyelse: Etapp 2, 2000–2003" (2001), 51, 57. Avesta municipality.
71. Avesta municipal council. Recording of meeting (March 20, 2003). Avesta municipality.

NOTES 201

72. Everbrand, interview; Karin Perers, interview, Avesta, April 27, 2006.
73. Åke Johansson, interview; Ulf Berg, interview, Avesta, May 17, 2006; Everbrand, interview; Linder, interview.
74. Kjersti Bosdotter, cited in Jan af Geijerstam, "Sammanfattning och reflexioner—referat av diskussioner," in *Industriarv i förändring: Rapport från en konferens, Koppardalen, Avesta, 7–9 mars 2006*, ed. Jan af Geijerstam (Avesta, 2007), 153.
75. Michel Rautenberg, "Industrial heritage, regeneration of cities and public policies in the 1990s: Elements of a French/British comparison," *International Journal of Heritage Studies* 18, no. 5 (2012); see also Anders Högberg, "The process of transformation of industrial heritage: Strengths and weaknesses," *Museum International* 63, no. 1–2 (2011), 34, 40.
76. Isacson, *Industrisamhället Sverige*, 240, 47.
77. Berg, interview; see also Everbrand, interview.
78. See, for example, Burell, interview; Everbrand, interview.
79. Burell, interview.
80. Everbrand, interview; see also Linder, interview.
81. Anne-Marie Nenzell, "Bubbelbadad visent tar plats mitt i stan," *Avesta Tidning* (October 17, 2001).
82. Berg, interview; Anders Hansson, interview, Avesta, September 18, 2003.
83. *Attraktiva Avesta*" (2002), 2. Avesta municipality.
84. Storm, *Koppardalen*, 84.
85. Forsberg, "Reproduktionen av den patriarkala bruksandan."
86. Göran Rydén, "Männens Bergslagen och kvinnornas," in *Bruksandan—hinder eller möjlighet? Rapport från en seminarieserie i Bergslagen. Ekomuseum Bergslagens skriftserie no 1*, ed. Ewa Bergdahl, Maths Isacson, and Barbro Mellander (Smedjebacken: Ekomuseum Bergslagen, 1997).

Bibliography

Archives and Unpublished Sources

Archives

ATA, the Swedish National Heritage Board Archives
Avesta municipal archives
Kävlinge municipal archives
National Museum of Denmark archives
Nordstjernan Corporation central archives, Engelsbergsarkivet

Private Collections

Morger, Kersti
Nisser, Marie
Perers, Karin
Steen Nielsen, Jørgen

Working Material

Avesta municipality
Dalarna County Administrative Board
Gällivare municipality

Interviews

Berg, Ulf. Avesta, May 17, 2006.
Burell, Jan. Avesta, April 26, 2006.
Carlson, Åsa. Malmö, November 4, 2010.
Daléus, Lennart. Saltsjöbaden, October 5, 2010.
Degutis, Algimantas (with Alfredas Jomantas). Vilnius, November 26, 2010.
Ebert, Wolfgang. Geldern, February 12, 2007.
Everbrand, Lars Åke. Avesta, April 16, 2003, and May 18, 2006.
Hansson, Anders. Avesta, September 18, 2003.

Johanson, Gunnar. Stockholm, August 14, 2012.
Johansson, Lennart. Malmberget, June 13, 2007.
Johansson, Åke. Avesta, May 18, 2006.
Linder, Kenneth. Avesta, May 18, 2006.
Lorentz, Håkan. Barsebäck, November 3, 2010.
Malmsten, Bo. Flemingsberg, October 7, 2010.
Nistad, Jan. Stockholm, October 19, 2010.
Nyström, Tommy. Malmberget, June 15, 2007.
Palmqvist, Roland. Kävlinge, October 31, 2011.
Perers, Karin. Avesta, April 27, 2006.
Pettersson, Per-Erik. Avesta, May 17, 2006.
Ševaldin, Viktor. Ignalina, November 25, 2010.
Steen Nielsen, Jørgen. Köpenhamn, November 2, 2010.
Taranger, Maria. Barsebäck, November 3, 2010.
Thamsten, Jan. Avesta, September 16, 2003.
Thelin, Lennart. Malmberget, June 13, 2007.
Urbonavi⊠ius, Saulius. Ignalina, November 25, 2010.
Vaišvila, Zigmas. Vilnius, November 24, 2010.
Öst, Leif. Barsebäck, November 3, 2010.

Websites

Cyber City web encyclopedia, http://urbanhistory.historia.su.se/cybercity/ (accessed November 12 and November 17, 2013).
Das Paradies, www.euromuse.net/en/exhibitions/exhibition/view-e/das-paradies/ (accessed August 29, 2013).
Encyclopedia Britannica, www.britannica.com (accessed September 25, 2013).
Energibevægelsen OOA. Organisationen til Oplysning om Atomkraft, www.ooa.dk (accessed June 19, 2012).
Environmental Protection Agency, International Brownfields Case Study, www.epa.gov (accessed October 15, 2007).
Landschaftspark Duisburg-Nord, www.landschaftspark.de (accessed March 19, 2007, and September 25, 2013).
Latz und Partner, www.latzundpartner.de/projects/detail/18 (accessed September 25, 2013).
Nara Document on Authenticity, www.iicc.org.cn/Info.aspx?ModelId=1&Id=422 (accessed April 12, 2014).
Oxford English Dictionary, www.oed.com (accessed September 25, 2013).
Sovietologist blog, http://sovietologist.blogspot.fr/2008/04/rbmk-reactors-and-weapons-grade.html (accessed April 15, 2014).
Universität Oldenburg, landscape ecology, www.landeco.uni-oldenburg.de/20716.html (accessed September 25, 2013).
Utøya memorial, www.bustler.net/index.php/article/swedish_artist_jonas_dahlberg_to_design_july_22_memorial_sites_in_norway/ (accessed April 9, 2014).

Vietnam Veterans Memorial, www.nybooks.com/articles/archives/2000/ nov/02/making-the-memorial/?page=1 (accessed April 12, 2014).
Völklinger Hütte, www.voelklinger-huette.org/de/faszination-weltkulturerbe/das-paradies/ (accessed August 29, 2013).

Other Unpublished Sources

Andersson, Gudrun. *Kvinnor och män i Malmbergets kåkstad: Med kåkstaden i Kiruna som referenspunkt.* D-uppsats. Luleå Tekniska Universitet, 2004.
Bohlin, Anna. "Land restitution and reconciliation: A lost opportunity? Emotion, land and heritage in post-apartheid South Africa." Paper presented at the healing-heritage workshop, University of Gothenburg, 2012.
Brown, Kate. "In the house that plutonium built: The history of plutonium, radiation and the communities that learned to love their bomb." Unpublished.
Butcher, P., P. Bystedt, M. Chouha, E. Uspuras, J. P. Weber, and K. Zilys. "TSO assistance towards the improvement of nuclear safety in Lithuania. Achivements and perspectives." Paper presented at the Eurosafe Forum 2001, Paris, France, 2001.
Eklund, Magnus. *Att möta krisen: Strukturomvandlingen i Avesta 1960–90 och kommunalt motagerande 1978–86.* C-uppsats. Uppsala Universitet, 2000.
Grahn, Niclas. *Barsebäcksverkets lokalisering och nedläggning: Hur förutsättningar och omständigheter för ett kärnkraftverk kan komma att förändras.* Arbetsrapporter. Kulturgeografiska institutionen nr. 710. Uppsala: Uppsala Universitet, 2010.
Gustafsson, Fredrik, and Pär Isling. *Gropen som svalde ett samhälle: Gropen i media 1956–1979.* C-uppsats. Luleå Tekniska Universitet, 2005.
Gylys, Jonas, and Leonas Ašmantas. "The specific nuclear energy problems in Lithuania." Paper presented at the 19th World Energy Congress, Sydney, Australia, 2004.
Hiller, Johan, Martii Laaksonen, Frida Munktell, Anna Roos, and Charlotta Thuresson. "Avesta projekt Kopparlänken dialog." KTH, 1999.
Olsson, Krister, and Marcus Adolphson. "Stadsstruktur, kulturvärden och identitet. Framtida flytt av Kiruna stad." Stockholm: KTH, 2008.
Tholander, Caroline. *Koppardalsprojektet ur ett uthållighetsperspektiv: Begrepps- och verktygslåda. Teorier och erfarenheter från Chalmers, m. fl.* Uppsats inom kursen "Stadsbyggnad med miljöhänsyn." Göteborg: CTH, 1999.
Wendland, Anna Veronika. "Inventing the atomograd. Nuclear as a way of life in Eastern Europe before and after Chernobyl." Unpublished.
Wik, Tina. "Bosnia-Herzegovina, Restoring war damaged built heritage in Bosnia-Herzegovina" (Conference report Bhopal, 2011). Available at http://tinawikarkitekter.se/publikationer/ (accessed April 9, 2014).

Published Sources and Literature

Newspapers

Arbetet
Avesta Tidning
Berlingske tidene
Dagens Nyheter
Dala-Demokraten
Expressen
Extra bladet
Falu Kuriren
Hallandsposten
Helsingborgs Dagblad
Information
Kristianstadsbladet
Norra Skåne
Norrländska Socialdemokraten
Nyheterna
Politiken
Skånska Dagbladet
Svenska Dagbladet
Sydsvenska Dagbladet
Östgöta Correspondenten
Östra Småland

Other Published Sources and Literature

Alzén, Annika. *Fabriken som kulturarv: Frågan om industrilandskapets bevarande i Norrköping 1950–1985*. Stockholm/Stehag: Brutus Östlings Bokförlag Symposion, 1996.

———. "Kulturarv i rörelse: En jämförande studie." In *Kulturarvens gränser: Komparativa perspektiv*, edited by Peter Aronsson, Bjarne Hodne, Birgitta Skarin Frykman, and John Ødemark. Göteborg: Bokförlaget Arkipelag, 2005.

Andersen, Håkon W., Terje Borgersen, Thomas Brandt, Knut Ove Eliassen, Ola Svein Stugu, and Audun Øfsti. *Fabrikken*. Oslo: Scandinavian Academic Press/Spartacus Forlag, 2004.

Andersson, Caj. "Om vi kände skräck för kärnkraften skulle vi inte arbeta här!" *Året runt* 46 (November 12, 1979).

Anshelm, Jonas. *Bergsäkert eller våghalsigt? Frågan om kärnavfallets hantering i det offentliga samtalet i Sverige 1950–2002*. Lund: Arkiv, 2006.

———. *Mellan frälsning och domedag: Om kärnkraftens politiska idéhistoria i Sverige 1945–1999*. Eslöv: Brutus Östlings Bokförlag Symposion, 2000.

Antrop, Marc. "A brief history of landscape research." In *The Routledge companion to landscape studies*, edited by Peter Howard, Ian Thompson, and Emma Waterton. 12–22. Abingdon: Routledge, 2013.

Aronsson, Peter. *Historiebruk: Att använda det förflutna.* Lund: Studentlitteratur, 2004.
Ashworth, Gregory, and Peter Howard. *European heritage planning and management.* Exeter: Intellect Ltd, 1999.
Avango, Dag. *Sveagruvan: Svensk gruvhantering mellan industri, diplomati och geovetenskap 1910–1934.* Stockholm papers in history and philosophy of technology, Trita-HOT: 2048. Stockholm: Jernkontoret, 2005.
Bahr, Betsy. "Adapting a future from the past: Reusing old industrial buildings for new industrial uses." In *Industrial heritage '84 proceedings: The fifth international conference on the conservation of the industrial heritage.* 191–95. Washington DC, 1984.
Bandolin, Gunilla, and Sverker Sörlin. *Laddade landskap—värdering och gestaltning av teknologiskt sublima platser, R-07–14.* Stockholm: SKB, 2007.
Bauer, Susanne, Karena Kalmbach, and Tatiana Kasperski. "From Pripyat to Paris, from grassroots memories to globalized knowledge production: The politics of Chernobyl fallout." In *Nuclear Portraits*, edited by Laurel MacDowell. Toronto: University of Toronto Press, forthcoming.
Bauman, Zygmunt. "From pilgrim to tourist—or a short history of identity." In *Questions of cultural identity*, edited by Stuart Hall and Paul du Gay. 18–36. London: SAGE Publications Ltd, 1996.
Beck, Ulrich. *World at risk.* Cambridge: Polity Press, 2009.
Bell, Daniel. *The coming of post-industrial society: A venture in social forecasting.* Harmondsworth: Penguin, 1973.
Bergdahl, Ewa, Maths Isacson, and Barbro Mellander, eds. *Bruksandan— hinder eller möjlighet? Rapport från en seminarieserie i Bergslagen*, Ekomuseum Bergslagens skriftserie no 1. Smedjebacken: Ekomuseum Bergslagen, 1997.
Bergverksstatistik 2012. Uppsala: SGU, 2013.
Bjurling, Oscar. *Sydkraft—samhälle: En berättelse i text och bild.* Malmö: Sydkraft, 1982.
Björck, Björn, Birgitta Sundberger, Kersti Lenngren, and Mille Törnblom. *Koppardalens förnyelse, etapp 2. Projekt dokumentation: Identifiering av värdebärare. Koppardalen i förändring.* Avesta: Avesta kommun, 2001.
Blaser, Mario. *Storytelling globalization from the Chaco and beyond.* Durham: Duke University Press, 2010.
Boëthius, Maria-Pia. *Heder och samvete: Sverige och andra världskriget.* Stockholm: Ordfront, 1999.
Bordage, Fazette, ed. *The factories: Conversions for urban culture.* Basel: Birkhäuser, 2002.
Borg, Henrik. "Barsebäck, en historia om trivsel, säkerhet, demonstrationer och kulturarv." In *Kärnkraft retro*, edited by Jan Garnert. Dædalus, Tekniska museets årsbok, 139–47. Stockholm: Tekniska museet, 2008.
Borg, Henrik, and Helen Sannerstedt. *Barsebäcks kärnkraftverk. Rapport 2006:57.* Regionmuseet Kristianstad/Landsantikvarien i Skåne, 2006.

Bourke, Joanna. *Fear: A cultural history.* London: Virago, 2005.

Brown, Kate. "Gridded lives: Why Kazakhstan and Montana are nearly the same place." *The American Historical Review* 106, no. 1 (2001): 17–48.

———. *Plutopia: Nuclear families, atomic cities, and the great Soviet and American plutonium disasters.* Oxford: Oxford University Press, 2013.

Brunnström, Lasse. *Kiruna—ett samhällsbygge i sekelskiftets Sverige.* Umeå: Umeå Universitet, diss., 1981.

Buchanan, R. Angus. *Industrial archaeology in Britain.* Harmondsworth: Penguin Books Ltd, 1972.

Bursell, Barbro. "Bruket och bruksandan." In *Bruksandan—hinder eller möjlighet? Rapport från en seminarieserie i Bergslagen. Ekomuseum Bergslagens skriftserie no 1*, edited by Ewa Bergdahl, Maths Isacson, and Barbro Mellander. 10–20. Smedjebacken: Ekomuseum Bergslagen, 1997.

Buzar, Stefan. *Energy poverty in Eastern Europe: Hidden geographies of deprivation.* Aldershot: Ashgate, 2007.

Cheape, Hugh, Mary-Cate Garden, and Fiona McLean. "Editorial: Heritage and the environment." *International Journal of Heritage Studies* 15, no. 2–3 (2009): 104–107.

Cinis, Andis, Marija Drėmaitė, and Mart Kalm. "Perfect representations of Soviet planned space: Mono-industrial towns in the Soviet Baltic Republics in the 1950s–1980s." *Scandinavian Journal of History* 33, no. 3 (2008): 226–46.

Cossons, Neil. *The BP book of industrial archaeology.* Third ed. Newton Abbot Devon: David & Charles, 1993.

Cossons, Neil, and Kenneth Hudson, eds. *Industrial archaeologists' guide 1969–70.* Newton Abbot Devon, 1969.

Covo, Gaëlle. "Spatial planning, structural change and regional development policies within the Ruhr area in Germany." In *Exploring the Ruhr in Germany.* Bochum, 2001.

"'Cultural' industrial plant engineer KRESTA: Stage setting for Bregenz Festival." *GAW group imteam: News from the group*, no. 1 (2005).

Danielsen, Oluf. *Atomkraften under pres. Dansk debat om atomkraft 1974–85.* Roskilde: Roskilde Universitetsforlag, 2006.

Davis, Peter. *Ecomuseums: A sense of place.* Leicester museum studies series. London: Leicester University Press, 1999.

Dawson, Jane I. *Eco-nationalism. Anti-nuclear activism and national identity in Russia, Lithuania, and Ukraine.* Durham: Duke University Press, 1996.

Dawson, Jane I., and Robert G. Darst. "Meeting the challenge of permanent nuclear waste disposal in an expanding Europe: Transparency, trust and democracy." *Environmental Politics* 15, no. 4 (2006): 610–27.

Delony, Eric N. "Industrial archeology in the United States 1981–1984." In *Industrial heritage '84 national reports: The fifth international conference on the conservation of the industrial heritage.* 117–23. Washington DC, 1984.

BIBLIOGRAPHY

Desfor, Gene, and John Jørgensen. "Flexible urban governance: The case of Copenhagen's recent waterfront development." *European planning studies* 12, no. 4 (2004): 479-96.

Desfor, Gene, and Lucian Vesalon. "Urban expansion and industrial nature: A political ecology of Toronto's port industrial district." *International Journal of Urban and Regional Research* 32, no. 3 (2008): 586-603.

Dettmar, Jörg. "Forests for shrinking cities? The project 'Industrial Forests of the Ruhr.'" In *Wild urban woodlands: New perspectives for urban forestry*, edited by Ingo Kowarik and Stefan Körner. 263-76. Berlin, Heidelberg: Springer-Verlag Berlin Heidelberg, 2005.

Dicks, Bella. *Heritage, place, and community*. Cardiff: University of Wales Press, 2000.

Drėmaitė, Marija. "Industrial heritage in a rural country: Interpreting the industrial past in Lithuania." In *Industrial heritage around the Baltic Sea*, edited by Marie Nisser, Maths Isacson, Anders Lundgren, and Andis Cinis. Uppsala Studies in Economic History 92, 65-78. Uppsala: Acta Universitatis Upsaliensis, 2012.

Edensor, Tim. *Industrial ruins: Spaces, aesthetics, and materiality*. Oxford: Berg, 2005.

En alternativ energiplan for Danmark. København: OOA, 1983.

Endres, Danielle. "Sacred land or national sacrifice zone: The role of values in the Yucca Mountain participation process." *Environmental Communication* 6, no. 3 (September 1, 2012): 328-45.

Everbrand, Lars Åke, and Dan Ola Norberg. *Koppardalsprojektet inför etapp 2*. Avesta, 2000.

Feldmann, Beate. "'The Pit' and 'the ghetto.' On heritage, identity and generation." In *Malmberget. Structural change and cultural heritage processes. A case study*, edited by Birgitta Svensson and Ola Wetterberg. 28-41. Stockholm: The Swedish National Heritage Board, 2009.

Feldmann Eellend, Beate. *Visionära planer och vardagliga praktiker: Postmilitära landskap i Östersjöområdet*. Stockholm Studies in Ethnology 7. Stockholm: Acta Universitatis Stockholmiensis, diss., 2013.

Fjæstad, Maja. "Demokratins triumf eller fiasko? Folkomröstningen om kärnkraft i retrospektiv." In *Kärnkraft retro*, edited by Jan Garnert. Dædalus, Tekniska museets årsbok. 65-75. Stockholm: Tekniska museet, 2008.

———. "Ett kärnkraftverk återuppstår: Från SNR300 till Wunderland Kalkar." *Bebyggelsehistorisk tidskrift*, no. 63 (2012): 26-38.

Foote, Kenneth E. *Shadowed ground: America's landscapes of violence and tragedy*, revised edition. Austin: University of Texas Press, 2003.

Forsberg, Gunnel. "Reproduktionen av den patriarkala bruksandan." In *Bruksandan—hinder eller möjlighet? Rapport från en seminarieserie i Bergslagen. Ekomuseum Bergslagens skriftserie no 1*, edited by Ewa Bergdahl, Maths Isacson, and Barbro Mellander. 59-64. Smedjebacken: Ekomuseum Bergslagen, 1997.

Forsström, Gösta. *Gällivare kommun. Del 1. Malmberget. Malmbrytning och bebyggelse*. Luleå: Norrbottens Museum, 1973.
Forsström, Gösta, and Bo Strand. *Gällivare kommun. Del 2. Gällivare. Tätort och landsbygd*. Luleå: Norrbottens Museum, 1977.
Foss, Nicole. *Nuclear safety and international governance: Russia and Eastern Europe*. Oxford: Oxford Institute for Energy Studies, 1999.
Fox, Julia. "Mountaintop removal in West Virginia: An environmental sacrifice zone." *Organization & Environment*, no. 12 (1999): 163–83.
Fredriksson, Jan. "Alvar Aalto och den uteblivna förnyelsen av Avesta stadscentrum." *Dalarna* (1997): 141–54.
Fröhlig, Florence. *Painful legacy of World War II: Nazi forced enlistment. Alsatian/Mosellan prisoners of war and the Soviet prison camp of Tambov*. Stockholm Studies in Ethnology 8. Stockholm: Acta Universitatis Stockholmiensis, diss., 2013.
Föhl, Axel. "'Ein Abwurfplatz für Meteoriten.' Industriedenkmale und die IBA." *Industrie-Kultur* 3 (1999): 2–3.
Fördjupad översiktsplan för tätorten Gällivare-Malmberget-Koskullskulle. Gällivare kommun, 2003.
Gandy, Matthew. "Urban nature and the ecological imaginary." In *In the nature of cities: Urban political ecology and the politics of urban metabolism*, edited by Nikolas C. Heynen, Maria Kaika, and Erik Swyngedouw. 63–74. New York, NY: Routledge, 2006.
Ganser, Karl. "The present state of affairs…Karl Ganser in a conversation with Carl Steckeweh and Kunibert Wachten." In *Change without growth? Sustainable urban development for the 20th century. VI. Architecture biennale Venice 1996*, edited by Kunibert Wachten. 12–23. Braunschweig/Wiesbaden: Friedr. Vieweg & Sohn Verlagsgesellschaft mbH, 1996.
Garnert, Jan, ed. *Kärnkraft retro*. Stockholm: Tekniska museet, 2008.
Geijerstam, Jan af. *Landscapes of technology transfer: Swedish ironmakers in India 1860–1864*. Jernkontorets bergshistoriska skriftserie: 42. Stockholm papers in the history of philosophy of technology, Trita-HOT: 2045. Stockholm: Jernkontoret, 2004.
———. "Sammanfattning och reflexioner—referat av diskussioner." In *Industriarv i förändring: Rapport från en konferens, Koppardalen, Avesta, 7–9 mars 2006*, edited by Jan af Geijerstam. 146–53. Avesta, 2007.
Gelotte, Hanna, Eva Dahlström Rittsél, and Anna Ulfstrand. *Nästa hållplats Södertälje*. Stockholm: Länsstyrelsen i Stockholms län, 2006.
Giblin, John Daniel. "Post-conflict heritage: symbolic healing and cultural renewal." *International Journal of Heritage Studies* (2013): 1–19.
Gibson, Lisanne, and John Pendlebury, eds. *Valuing historic environments*. Farnham: Ashgate, 2009.
Gineitiene, Dalia, Erika Jörgensen, Mikael Klintman, and Leonardas Rinkevičius. "Public risk perceptions of nuclear power. The case of Sweden and Lithuania." In *Social processes and the environment. Lithuania and Sweden*, edited by Anna-Lisa Lindén and Leonardas Rinkevičius. 121–64. Lund: Lund University, 1999.

BIBLIOGRAPHY

Gjestrum, John Aage. "En bibliografi om økomuseer." *Nordisk Museologi*, no. 2 (1996): 57–70.
Graham, Brian J., and Peter Howard, eds. *The Ashgate research companion to heritage and identity*. Aldershot: Ashgate, 2008.
Greider, Göran. *När fabrikerna tystnar: Dikter*. Stockholm: Bonnier, 1995.
Grime, J. Philip, and Simon Pierce. *The evolutionary strategies that shape ecosystems*. John Wiley & Sons, Ltd, 2012.
Gustafsson, Bo, ed. *Post-industrial society: Proceedings of an international symposium held in Uppsala from 22 to 25 March 1977 to mark the occasion of the 500th anniversary of Uppsala university*. London: Croom Helm, 1979.
Gällivare generalplan. Stockholm: Eglers stadsplanebyrå, 1967.
Hagerman, Chris. "Shaping neighborhoods and nature: Urban political ecologies of urban waterfront transformations in Portland, Oregon." *Cities* 24, no. 4 (2007): 285–97.
Hansson, Staffan. "Malm, räls och elektricitet: Skapandet av ett teknologiskt megasystem i Norrbotten 1880–1920." In *Den konstruerade världen: Tekniska system i historiskt perspektiv*, edited by Pär Blomkvist and Arne Kaijser. 45–76. Eslöv: Brutus Östlings bokförlag Symposion, 1998.
Harnow, Henrik. *Danmarks industrielle miljøer*. Odense: Syddansk Universitetsforlag, 2011.
Harvey, David. *The condition of postmodernity: An enquiry into the origins of cultural change*. Malden, MA: Blackwell Publishers Inc, 1990.
———. "Emerging landscapes of heritage." In *The Routledge companion to landscape studies*, edited by Peter Howard, Ian Thompson, and Emma Waterton. 152–65. Abingdon: Routledge, 2013.
Hecht, Gabrielle. *Being nuclear: Africans and the global uranium trade*. Cambridge, MA: MIT Press, 2012.
———, ed. *Entangled geographies: Empire and technopolitics in the global Cold War*. Cambridge, MA: MIT Press, 2011.
Hedlund, Tina, Peter Popper, and Peter Ögren. "Miljörapport 2005. LKAB Malmberget. Yttre miljö. Publicerad 2006-03-27." LKAB, 2006.
Hedskog, Bo. *Återanvändning av industri- och specialbyggnader: Fastighetsekonomiska, tekniska och funktionella aspekter på val av ny användning*. TRITA-FAE-1012 meddelande 5:12. Stockholm: Institutionen för fastighetsekonomi, KTH, 1982.
Hedström, Jan-Olof. "Hur kan det vara lagligt att flytta staden Kiruna för att ge plats åt gruvan?" *Plan*, no. 3 (2007): 30–31.
Heinemann, Ulrich. "Industriekultur: Vom Nutzen zum Nachteil für das Ruhrgebiet." *Forum Industriedenkmalpflege und Geschichtskultur* 1 (2003): 56–58.
Helsing Almaas, Ingerid. "Regenerating the Ruhr—IBA Emscher Park project for the regeneration of Germany's Ruhr region." *The Architectural Review* CCV, no. 1224 (1999): 13–14.
Henare, Amiria J. M., Martin Holbraad, and Sari Wastell, eds. *Thinking through things: Theorising artefacts ethnographically*. Milton Park, Abingdon, Oxon: Routledge, 2007.

Henriksson, Adolf. "Det moderna Malmberget." *LKAB-tidningen*, no. 1 (1963): 9.
Herttrich, Michael, Rolf Janke, and Peter Kelm. "International co-operation to promote nuclear-reactor safety in the former USSR and Eastern Europe." In *Green globe yearbook of international co-operation on environment and development 1994*, edited by Helge Ole Bergesen and Georg Parmann. 89–101. Oxford: Oxford University Press, 1994.
Hirdman, Yvonne, Jenny Björkman, and Urban Lundberg. *Sveriges historia. 1920–1965*. Stockholm: Norstedt, 2012.
Holmstedt, Erik, and Sverker Sörlin. *Inte längre mitt hem. Malmberget 1969–1978, 2007–2008*. Luleå: Black Island Books, 2008.
Holtorf, Cornelius, and Anders Högberg. "Heritage futures and the future of heritage." In *Counterpoint: Essays in archaeology and heritage studies in honour of professor Kristian Kristiansen*, edited by Sophie Bergerbrant and Serena Sabatini. 739–46. Oxford: Archaeopress, 2013.
Houltz, Anders. "Fabriken som aldrig blir färdig: Volvo Torslandaverken och 1960-talets svenska industrinationalism." *Fabrik og bolig*, no. 2012 (2012): 4–13.
———. *Teknikens tempel: Modernitet och industriarv på Göteborgsutställningen 1923*. Stockholm papers in the history and philosophy of technology, Trita-HOT: 2041. Hedemora: Gidlund, 2003.
Houston, Donna. "Environmental justice storytelling: Angels and isotopes at Yucca Mountain, Nevada." *Antipode* 45, no. 2 (2013): 417–35.
Hudson, Kenneth. "Ecomuseums become more realistic." *Nordisk Museologi*, no. 2 (1996): 11–19.
———. *Industrial archaeology: An introduction*. London: John Baker Publishers Ltd, 1963.
Huske, Joachim. *Die Steinkohlenzechen im Ruhrrevier. Daten und Fakten von den Anfängen bis 2005, 3. überarbeitete und erweiterte Auflage*. Bochum: Deutsches Bergbau-Museum, 2006. 3rd. 1987.
"Hyttan blev konsthall." *Attraktiva Avesta* (2002).
Höfer, Wolfram, and Vera Vicenzotti. "Post-industrial landscapes: Evolving concepts." In *The Routledge companion to landscape studies*, edited by Peter Howard, Ian Thompson and Emma Waterton. 405–16. Abingdon: Routledge, 2013.
Högberg, Anders. "The process of transformation of industrial heritage: Strengths and weaknesses." *Museum International* 63, no. 1–2 (2011): 34–42.
Högselius, Per. "Connecting East and West? Electricity systems in the Baltic region." In *Networking Europe. Transnational infrastructures and the shaping of Europe 1850–2000*, edited by Erik van der Vleuten and Arne Kaijser. 245–75. Sagamore Beach, MA: Science History Publications, 2006.
Högselius, Per, and Arne Kaijser. *När folkhemselen blev internationell: Elavregleringen i historiskt perspektiv*. Stockholm: SNS Förlag, 2007.

BIBLIOGRAPHY

Isacson, Maths. "Bruksandan—hinder eller möjlighet? Sammanfattning av fyra seminarier i Bergslagen 1995-1997." In *Bruksandan—hinder eller möjlighet? Rapport från en seminarieserie i Bergslagen*. *Ekomuseum Bergslagens skriftserie no 1*, edited by Ewa Bergdahl, Maths Isacson, and Barbro Mellander. 120-32. Smedjebacken: Ekomuseum Bergslagen, 1997.

———. *Industrisamhället Sverige: Arbete, ideal och kulturarv*. Lund: Studentlitteratur, 2007.

———. "Industrisamhällets faser och industriminnesforskningens uppgifter." In *Industrins avtryck: Perspektiv på ett forskningsfält*, edited by Dag Avango and Brita Lundström. Stockholm/Stehag: Symposion, 2003.

———. "Tre industriella revolutioner?" In *Industrialismens tid: Ekonomiskhistoriska perspektiv på svensk industriell omvandling under 200 år*, edited by Maths Isacson and Mats Morell. 11-28. Stockholm: SNS förlag, 2002.

Jauhiainen, Jussi S. "Waterfront redevelopment and urban policy: The case of Barcelona, Cardiff and Genoa." *European Planning Studies* 3, no. 1 (1995): 3-23.

Jensen, Ola W., and Håkan Karlsson. *Archaeological conditions: Examples of epistemology and ontology*. GOTARC. Serie C, Arkeologiska skrifter, 0282-9479. Göteborg: Göteborgs Universitet, 2000.

JM "Kvarnen: Warehouse living på Norra Älvstranden." Advertising brochure (2007).

Joelsson, Johan. "Kärnkraft som attraktion." *DIK-forum*, no. 6 (2010): 16-20.

Johansson, E., and M. Olsén. *Gällivares och Malmbergets födelse. En väggmålning av Erling Johansson i Centralskolan, Gällivare*. Kalix: Gellivare Sockens Hembygdsförening, undated.

Johansson, K. "Vid kanten av Gropen." In *Norrbottens museum. Människors upplevelser av att bo i Malmberget i dag. Norrbottenprojektet. En nutidsdokumentation 1994-1995*. Norrbottens Museum, 1996.

Jorgensen, Anna, and Richard Keenan, eds. *Urban wildscapes*. New York: Routledge, 2011.

Josephson, Paul R. *Industrialized nature: Brute force technology and the transformation of the natural world*. Washington DC: Island Press, 2002.

———. *Red atom: Russia's nuclear power program from Stalin to today*. New York: W. H. Freeman and Company, 1999.

———. "Technological utopianism in the twenty-first century. Russia's nuclear future." *History and technology* 19, no. 3 (2003): 277-92.

Jönsson, Lars-Eric, and Birgitta Svensson, eds. *I industrisamhällets slagskugga: Om problematiska kulturarv* (Stockholm: Carlsson, 2005).

Jörnmark, Jan. *Övergivna platser*. Lund: Historiska media, 2007.

Kaijser, Arne. "From tile stoves to nuclear plants—the history of Swedish energy systems." In *Building sustainable energy systems: Swedish experiences*, edited by Semida Silveira. 57-93. Stockholm: Svensk byggtjänst, 2001.

Kaijser, Arne. "Redirecting power: Swedish nuclear power policies in historical perspective." *Annual Review of Energy and the Environment* 17 (1992): 437–62.

———. "Trans-border integration of electricity and gas in the Nordic countries, 1915–1992." *Polhem*, no. 1 (1997): 4–43.

Kaijser, Arne, and Per Högselius. *Resurs eller avfall? Politiken kring hanteringen av använt kärnbränsle i Finland, Tyskland, Ryssland och Japan, R-07–37.* Stockholm: SKB, 2007.

Karaliūtė, Renata. "Nuclear knowledge management and preservation in Lithuania." *International Journal of Nuclear Knowledge Management* 1, no. 3 (2005): 217–22.

Karvonen, Andrew. *Politics of urban runoff: Nature, technology and the sustainable city.* Cambridge, MA: MIT Press, 2011.

Kasperski, Tatiana. "Chernobyl's aftermath in political symbols, monuments and rituals: Remembering the disaster in Belarus." *Anthropology of East Europe Review* 30, no. 1 (2012): 82–99.

Kattwinkel, Mira, Robert Biedermann, and Michael Kleyer. "Temporary conservation for urban biodiversity." *Biological Conservation*, no. 144 (2011): 2335–43.

Keil, Andreas. "Use and perception of post-industrial urban landscapes in the Ruhr." In *Wild urban woodlands: New perspectives for urban forestry*, edited by Ingo Kowarik and Stefan Körner. 117–30. Berlin, Heidelberg: Springer-Verlag Berlin Heidelberg, 2005.

Keil, Andreas, and Burkhard Wetterau. *Metropolis Ruhr: A regional study of the new Ruhr.* Essen: Regionalverband Ruhr, 2013.

Kidney, Walter C. *Working places: The adaptive use of industrial buildings.* Pittsburgh: Ober Park Associates, Inc., 1976.

Kift, Dagmar. "Heritage and history: Germany's industrial museums and the (re-) presentation of labour." *International Journal of Heritage Studies* 17, no. 4 (July 1, 2011): 380–89.

Kilper, Heiderose, and Gerald Wood. "Restructuring policies: The Emscher park international building exhibition." In *The rise of the rustbelt*, edited by Philip Cooke. 208–30. London: UCL Press, 1995.

Knudsen, Henrik. *Risøs reaktorer: Registrering og dokumentering af bevaringsværdige genstande fra Forskningscenter Risøs reaktorfaciliteter med henblik på at bevare Danmarks nukleare kulturarv.* Bjerringbro: Kulturarvsstyrelsen, 2006.

Kopustinskas, V., E. Urbonavičius, A. Kaliatka, S. Rimkevičius, E. Ušpuras, A. Bagdonas, P. Hellström, and G. Johanson. "An approach to estimate radioactive release frequency from Ignalina RBMK-1500 reactor in Lithuania." *Zagadnienia eksploatacji maszyn* 1, no. 149 (2007): 183–99.

Kowarik, Ingo. "Novel urban ecosystems, biodiversity, and conservation." *Environmental Pollution*, no. 159 (2011): 1974–83.

———. "Wild urban woodlands: Towards a conceptual framework." In *Wild urban woodlands: New perspectives for urban forestry*, edited by Ingo

BIBLIOGRAPHY 215

Kowarik and Stefan Körner. 1–32. Berlin, Heidelberg: Springer-Verlag Berlin Heidelberg, 2005.

Kowarik, Ingo, and Andreas Langer. "Natur-Park Südgelände: Linking conservation and recreation in an abandoned railyard in Berlin." In *Wild urban woodlands: New perspectives for urban forestry*, edited by Ingo Kowarik and Stefan Körner. 287–99. Berlin, Heidelberg: Springer-Verlag Berlin Heidelberg, 2005.

Krohn Andersson, Fredrik. "Kärnkraftens arkitektur." In *Kärnkraft retro*, edited by Jan Garnert. Dædalus, Tekniska museets årsbok, 105–21. Stockholm: Tekniska museet, 2008.

———. *Kärnkraftverkets poetik: Begreppsliggöranden av svenska kärnkraftverk 1965–1973*. Stockholm: Stockholms universitet, diss., 2012.

Krupar, Shiloh R. *Hot spotter's report: Military fables of toxic waste*. Minneapolis: University of Minnesota Press, 2013.

Kumar, Krishan. *Prophecy and progress: The sociology of industrial and post-industrial society*. Harmondsworth: Penguin, 1978.

Kverndokk, Kyrre. *Pilegrim, turist og elev: Norske skoleturer til døds- og konsentrasjonsleirer*. Linköping: Linköpings Universitet, diss., 2007.

Kåks, Helena. *Avesta: Industriarbete och vardagsliv genom 400 år*. DFR-rapport 2002:1. Falun: Dalarnas forskningsråd, 2002.

Labanauskas, Liutauras. "Social aspects of the functioning of the Ignalina nuclear power plant." *Viešoji politika ir administravimas*, no. 22 (2007): 78–84.

Lachmund, Jens. *Greening Berlin: The co-production of science, politics, and urban nature*. Cambridge, MA: MIT Press, 2013.

Latz, Peter. "Landscape Park Duisburg-Nord: The metamorphosis of an industrial site." In *Manufactured sites: Rethinking the post-industrial landscape*, edited by Niall Kirkwood. 150–61. New York: Spon Press, 2001.

LeCain, Timothy J. *Mass destruction: The men and giant mines that wired America and scarred the planet*. New Brunswick, NJ: Rutgers University Press, 2009.

Lekan, Thomas, and Thomas Zeller. *Germany's nature: Cultural landscapes and environmental history*. New Brunswick, NJ: Rutgers University Press, 2005.

Lindqvist, Svante. *Än kan man köpa en Portello på Sporthallsfiket. Berättelser omkring Malmfälten*. Gällivare: Gellivare Sockens Hembygdsförening, 1995.

Lindqvist, Sven. *Gräv där du står: Hur man utforskar ett jobb*. Stockholm, 1978.

Lindström, Stefan. "Implementing the welfare state: The emergence of Swedish atomic energy research policy." In *Center on the periphery: Historical aspects of 20th-century Swedish physics*, edited by Svante Lindqvist, Marika Hedin, and Thomas Kaiserfeld. 179–95. Canton, MA: Science History Publications, 1993.

Linné, Carl-Erik. "Kommun och industri i samarbete." *LKAB-tidningen*, no. 1 (1964): 20–24.
Lipset, Seymour Martin, ed. *The third century: America as a post-industrial society*. Stanford University: Hoover Institution Press, 1980 [1979].
Littke, Göran. *Malmberget 1888–1963*. Malmberget: LKAB, 1963.
Ljunggren, Sven. "Undersökningen av kaptensmalmen under Malmbergets samhälle." *LKAB-tidningen*, no. 2 (1959): 8–9.
"LKAB 1.10.1957–30.9.1967." *LKAB-tidningen*, no. 3 (1967).
LKAB. *Årsredovisning*. Luleå: LKAB, 2006.
Logan, William Stewart, and Keir Reeves, eds. *Places of pain and shame: Dealing with "difficult heritage."* Abingdon: Routledge, 2009.
Lowenthal, David. *The heritage crusade and the spoils of history*. Cambridge: Cambridge University Press, 1998.
———. *The past is a foreign country*. Cambridge: Cambridge University Press, 1985.
———. "Restoration: Synoptic reflections." In *Envisioning landscapes, making worlds: Geography and the humanities*, edited by Stephen Daniels. 209–26. Milton Park, Abingdon, Oxon: Routledge, 2011.
———. "Stewarding the future." *CRM: The Journal of Heritage Stewardship*, no. 2 (2005).
Lundberg, Arne S. "Malmbergets framtid." *LKAB-tidningen*, no. 3 (1961): 3.
Löfström, Bernt. "Västtyskland—vår största malmkund." *LKAB-tidningen*, no. 3 (1967).
Lövgren, Einar. *Folkare din hembygd*. Avesta, 1977.
Macdonald, Sharon. *Difficult heritage: Negotiating the Nazi past in Nuremberg and beyond*. London: Routledge, 2009.
Martins Holmberg, Ingrid. "The historicisation of Malmberget." In *Malmberget. Structural change and cultural heritage processes. A case study*, edited by Birgitta Svensson and Ola Wetterberg. 42–54. Stockholm: The Swedish National Heritage Board, 2009.
Masco, Joseph. *The nuclear borderlands: The Manhattan project in post–Cold War New Mexico*. Princeton, NJ: Princeton University Press, 2006.
Massey, Doreen. "Places and their pasts." *History Workshop Journal* 39 (1995): 182–92.
Masuda, Yoneji. *The information society: As post-industrial society*. Tokyo: Institute for the Information Society, 1980.
Maure, Marc. "Identitet, økologi, deltakelse: Om museenes nye rolle." In *Økomuseumsboka: Identitet, økologi, deltakelse*, edited by John Aage Gjestrum and Marc Maure. 16–32. Tromsø: Norsk ICOM, 1988.
Millington, Nate. "Post-industrial imaginaries: Nature, representation and ruin in Detroit, Michigan." *International Journal of Urban and Regional Research* 37, no. 1 (2013): 279–96.
Moberg, Åsa. *Ett extremt dyrt och livsfarligt sätt att värma vatten: En bok om kärnkraft*. Stockholm: Natur och kultur, 2014.

Moreira De Marchi, Polise. "Ruhrgebiet: redesigning an industrial region." In *Exploring the Ruhr in Germany*. Bochum, 2001.

Müller, Rita. "Museums designing for the future: Some perspectives confronting German technical and industrial museums in the twenty-first century." *International Journal of Heritage Studies* 19, no. 5 (July 1, 2013): 511–28.

"Mysterierna i Malmberget." *LKAB-tidningen*, no. 1 (1962).

Nielsen, Henry, and Henrik Knudsen. "The troublesome life of peaceful atoms in Denmark." *History and Technology* 26, no. 2 (2010): 91–118.

Ninjalicious. *Access all areas: A users's guide to the art of urban exploration*. Toronto: Infiltration, 2005.

Nisser, Marie. "Industriminnen på den internationella arenan." In *Industrisamhällets kulturarv: Betänkande av Delegationen för industrisamhällets kulturarv*. 217–26. Stockholm: Statens offentliga utredningar (SOU) 2002:67, 2002.

———. "Industriminnen under hundra år." *Nordisk Museologi*, no. 1 (1996): 73–82.

———. *Industriminnen: En bok om industri- och teknikhistoriska bebyggelsemiljöer*. Stockholm, 1979.

———. "Nytt liv i Europas gamla industriområden." *Kulturmiljövård*, no. 6 (1994).

Nisser, Marie, and Fredric Bedoire. "Industrial monuments in Sweden: Conservation, documentation and research." In *The industrial heritage: The third international conference on the conservation of industrial monuments. Transactions 2. Scandinavian reports*, edited by Marie Nisser. 73–78. Stockholm: Nordiska museet, 1978.

Nordlund, Therese. *Att leda storföretag: En studie av social kompetens och entreprenörskap i näringslivet med fokus på Axel Ax:son Johnson och J. Sigfrid Edström, 1900–1950*. Stockholm studies in economic history. Stockholm: Acta Universitatis Stockholmiensis, 2005.

Nye, David E. *Technology matters: Questions to live with*. Cambridge: MIT Press, 2006.

Nye, David E., and Sarah Elkind, eds. *The Anti-landscape*. Amsterdam: Rodopi, 2014.

Nylund, Thomas. "Kiruna—att plan era för stadsflytt." *Plan*, no. 3 (2007): 20–27.

Olshammar, Gabriella. "Ruin landscape: A problem or history?" In *Malmberget. Structural change and cultural heritage processes. A case study*, edited by Birgitta Svensson and Ola Wetterberg. 15–27. Stockholm: The Swedish National Heritage Board, 2009.

Olwig, Kenneth Robert. *Landscape, nature, and the body politic: from Britain's renaissance to America's new world*. Madison: University of Wisconsin Press, 2002.

Orrje, Henrik. *Konsten att gestalta offentliga miljöer: Samverkan i tanke och handling*. Stockholm: Statens konstråd, 2013.

Papayannis, Thymio, and Peter Howard. "Editorial: Nature as heritage." *International Journal of Heritage Studies* 13, no. 4–5 (2007): 298–307.

Parr, Joy. *Sensing changes: Technologies, environments, and the everyday, 1953–2003*. Vancouver: UBC Press, 2010.

Perers, Karin. "Avesta Art." *Folkarebygden* (1995).

Perrow, Charles. *Normal accidents: Living with high-risk technologies*. Princeton, NJ: Princeton University Press, 1999.

Petryna, Adriana. *Life exposed: Biological citizens after Chernobyl*. New Jersey: Princeton University Press, 2002.

Pettersson, Richard. "Plats, ting och tid: Om autenticitetsuppfattning utifrån exemplet Rom." In *Topos: Essäer om tänkvärda platser och platsbundna tankar*. 202–16. Stockholm: Carlssons Bokförlag, 2006.

Plieninger, Tobias, and Claudia Bieling. "Resilience and cultural landscapes: Opportunities, relevance and ways ahead." In *Resilience and the cultural landscape: Understanding and managing change in human-shaped environments*, edited by Tobias Plieninger and Claudia Bieling. 328–42. Cambridge: Cambridge University Press, 2012.

Qviström, Mattias. "Network ruins and green structure development: An attempt to trace relational spaces of a railway ruin." *Landscape Research* 37, no. 3 (2012): 257–75.

Qviström, Mattias, and Katarina Saltzman. "Exploring landscape dynamics at the edge of the city: Spatial plans and everyday places at the inner urban fringe of Malmö, Sweden." *Landscape Research* 31, no. 1 (2006): 21–41.

Raines, Anne Brownley. "Wandel durch (Industrie) Kultur [Change through (industrial) culture]: Conservation and renewal in the Ruhrgebiet." *Planning Perspectives* 26, no. 2 (2011): 183–207.

Rautenberg, Michel. "Industrial heritage, regeneration of cities and public policies in the 1990s: Elements of a French/British comparison." *International Journal of Heritage Studies* 18, no. 5 (2012): 513–25.

Regionmuseet. *Stort, fult, farligt? Barsebäcks kärnkraftverk och kulturarvet. Rapport 2003:115*. Regionmuseet Kristianstad/Landsantikvarien i Skåne, 2003.

Ricœur, Paul. *Memory, history, forgetting*. Translated by Kathleen Blamey and David Pellauer. Chicago: University of Chicago Press, 2004.

Riesto, Svava. *Digging Carlsberg: Landscape biography of an industrial site undergoing redevelopment*. Forest & Landscape: University of Copenhagen, 2011.

Rinkevi⊠ius, Leonardas. "Attitudes and values of the Lithuanian green movement in the period of transition." *Filosofija, sociologija*, no. 1 (2001): 72–79.

Rix, Michael. "Industrial archaeology." *The Amateur Historian* 2, no. 8 (1955): 225–29.

Robertson, Iain J. M., ed. *Heritage from below*. Farnham: Ashgate Pub. Company, 2012.

Rodell, Magnus. "Monumentet på gränsen: Om den rumsliga vändningen och ett fredsmonument." *Scandia* 74, no. 2 (2008): 15–51.

Rudberg, Eva. *Alvar Aalto i Sverige*. Arkitekturmuseets skriftserie nr 12. Stockholm: Arkitekturmuseet, 2005.

Rydén, Göran. "Männens Bergslagen och kvinnornas." In *Bruksandan— hinder eller möjlighet? Rapport från en seminarieserie i Bergslagen. Ekomuseum Bergslagens skriftserie no 1*, edited by Ewa Bergdahl, Maths Isacson, and Barbro Mellander. 46–58. Smedjebacken: Ekomuseum Bergslagen, 1997.

Sandell, Klas. "Friluftsliv, platsperspektiv och samhällskritik." In *Topos: Essäer om tänkvärda platser och platsbundna tankar*. 274–91. Stockholm: Carlssons Bokförlag, 2006.

———. "Var ligger utomhus? Om nya arenor för natur- och landskapsrelationer." In *Utomhusdidaktik*, edited by Iann Lundegård, Per-Olof Wickman, and Ammi Wohlin. 151–70. Lund: Studentlitteratur, 2004.

Sanering efter industrinedläggningar: Betänkande av industrisaneringsutredningen. Stockholm: Statens offentliga utredningar (SOU) 1982:10, 1982.

Schmid, Sonja D. "Celebrating tomorrow today: The peaceful atom on display in the Soviet Union." *Social Studies of Science* 36, no. 3 (2006): 331–65.

Senn, Alfred Erich. *Gorbachev's failure in Lithuania*. New York: St. Martin's Press, 1995.

———. *Lithuania awakening*. Berkeley: University of California Press, 1990.

Shaw, Robert. "The International Building Exhibition (IBA) Emscher Park, Germany: A model for sustainable restructuring?" *European Planning Studies* 10, no. 1 (January 1, 2002): 77–97.

Sillén, Gunnar. *Stiga vi mot ljuset: Om dokumentation av industri- och arbetarminnen*. Stockholm, 1977.

Silvén, Eva, and Anders Björklund, eds. *Svåra saker: Ting och berättelser som upprör och berör*. Stockholm: Nordiska museets förlag, 2006.

Sjöholm, Jennie. *Heritagisation of built environments: A study of the urban transformation in Kiruna, Sweden*. Luleå University of Technology: Licentiate thesis, 2013.

Sjöholm, Jennie, and Kristina L. Nilsson. *Malmfältens kulturmiljöprocesser*. Luleå: Luleå tekniska universitet, 2011.

Sjölander-Lindqvist, Annelie, Anna Bohlin, and Petra Adolfsson. *Delaktighetens landskap: Tillgänglighet och inflytande inom kulturarvssektorn*. Stockholm: Riksantikvarieämbetet, 2010.

Šliavaitė, Kristina. *From pioneers to target group. Social change, ethnicity and memory in a Lithuanian nuclear power plant community*. Lund: Lunds universitet, 2005.

Slotta, Rainer. "Industrial archaeology in the Federal Republic of Germany." In *Industrial heritage '84 national reports: The fifth international conference on the conservation of the industrial heritage*. 37–45. Washington DC, 1984.

Slovic, Paul. *The perception of risk*. London: Earthscan Publications, 2000.

Smith, Laurajane. *Uses of heritage*. New York: Routledge, 2006.
Smith, Laurajane, Paul Shackel, and Gary Campbell, eds. *Heritage, labour, and the working classes*. Abingdon, Oxon: Routledge, 2011.
Soyez, Dietrich. "Das amerikanische Industriemuseum 'Sloss Furnaces'—ein Modell für das Saarland?" *Annales. Forschungsmagazin der Universität des Saarlandes*, no. 1 (1988): 59–68.
———. "Europeanizing industrial heritage in Europe: Addressing its transboundary and dark sides." *Geographische Zeitschrift* 91, no. 1 (2009): 43–55.
Steinmetz, George. "Detroit: A tale of two crises." *Environment and Planning D: Society and Space* 27, no. 5 (2009): 761–70.
Stone-Mediatore, Shari. *Reading across borders: Storytelling and knowledges of resistance*. New York: Palgrave Macmillan, 2003.
Storm, Anna. *Hope and rust: Reinterpreting the industrial place in the late 20th century*. Stockholm: KTH diss., 2008.
———. *Koppardalen: Om historiens plats i omvandlingen av ett industriområde*. Stockholm papers in the history and philosophy of technology, Stockholm: KTH lic., 2005.
Storm, Anna, and Krister Olsson. "The pit: Landscape scars as potential cultural tools." *International Journal of Heritage Studies* 19, no. 7 (2013): 692–708.
Stratton, Michael. "Understanding the potential: Location, configuration and conversion options." In *Industrial buildings: Conservation and regeneration*, edited by Michael Stratton. 30–46. London: E & FN Spon, 2000.
Sörlin, Sverker. "Friction in the field: Meanings of military landscapes," In *Militære landskap: festspillutstillingen 2008 = Military landscapes: Bergen international festival exhibition 2008*, ed. Ingrid Book. Bergen: Bergen kunsthall, 2008.
———. "The trading zone between articulation and preservation: Production of meaning in landscape history and the problems of heritage decision-making." In *Rational decision-making in the preservation of cultural property: Report of the 86th Dahlem workshop, Berlin, March 26–31, 2000*, edited by Norbert S. Baer and Folke Snickars. 47–59. Berlin: Dahlem University Press, 2001.
Tafvelin Heldner, Magdalena. "Strumpstickor och pingpongbollar: Med Tekniska museet i atomåldern." In *Kärnkraft retro*, edited by Jan Garnert. Dædalus, Tekniska museets årsbok, 48–63. Stockholm: Tekniska museet, 2008.
Tafvelin Heldner, Magdalena, Eva Dahlström Rittsél, and Per Lundgren. *Ågesta: Kärnkraft som kulturarv*. Stockholm: Tekniska museet, Stockholms läns museum, Länsstyrelsen i Stockholms län, 2008.
Tolia-Kelly, Divya P. "Landscape and memory." In *The Routledge companion to landscape studies*, edited by Peter Howard, Ian Thompson, and Emma Waterton. 322–34. Abingdon: Routledge, 2013.

Trinder, Barrie, ed. *The Blackwell encyclopedia of industrial archaeology.* Oxford, 1992.
Tuan, Yi-Fu. *Topophilia: A study of environmental perception, attitudes, and values.* New York: Columbia University Press, 1974.
Turtinen, Jan. *Världsarvets villkor: Intressen, förhandlingar och bruk i internationell politik.* Acta Universitatis Stockholmiensis, Stockholm studies in Ethnology 1. Stockholm: Stockholm University, 2006.
Urry, John. "The place of emotions within place." In *Emotional geographies,* edited by Joyce Davidson, Liz Bondi, and Mick Smith. 77–83. Aldershot: Ashgate, 2005.
Varine, Hugues de. "L´Ecomusée." *Gazette,* no. 11, 2 (1978): 28–40.
Weart, Spencer R. *The rise of nuclear fear.* Cambridge, MA: Harvard University Press, 2012.
Weiss, Joachim, Wolfgang Burghardt, Peter Gausmann, Rita Haag, Henning Haeupler, Michael Hamann, Bertram Leder, Annette Schulte, and Ingrid Stempelmann. "Nature returns to abandoned industrial land: Monitoring succession in urban-industrial woodlands in the German Ruhr." In *Wild urban woodlands: New perspectives for urban forestry,* edited by Ingo Kowarik and Stefan Körner. 143–62. Berlin, Heidelberg: Springer-Verlag Berlin Heidelberg, 2005.
Widmalm, Sven, and Hjalmar Fors, eds. *Artefakter: industrin, vetenskapen och de tekniska nätverken.* Hedemora: Gidlund, 2004.
Vikström, Eva. *Bruksandan och modernismen: Brukssamhälle och folkhemsbygge i Bergslagen 1935–1975.* Nordiska museets handlingar 126. Stockholm1998.
———. *Industrimiljöer på landsbygden: Översikt över kunskapsläget.* Stockholm: Riksantikvarieämbetet, 1994.
———. *Platsen, bruket och samhället: Tätortsbildning och arkitektur 1860–1970.* Stockholm: Statens råd för byggnadsforskning, 1991.
Wilhelmson, Anders. "Nya Kiruna." *Plan,* no. 3 (2007): 32–39.
Willim, Robert. *Industrial cool: Om postindustriella fabriker.* Lund: Lunds Universitet, 2008.
Woods, Lebbeus. *Radical reconstruction.* New York: Princeton Architectural Press, 1997.
Zonabend, Françoise. *The nuclear peninsula.* Translated by J. A.Underwood. Cambridge: Cambridge University Press, 1993.
Zukin, Sharon. *Landscapes of power: From Detroit to Disney World.* Berkeley: University of California Press, 1991. First Paperback Printing.
———. *Loft living: Culture and capital in urban change.* New Brunswick, NJ: Rutgers University Press, 1982.

Index

9/11, events of, (New York), 68
Aalto, Alvar, 129, 131, 138, 141
Ågesta nuclear power plant, 51, 70, 173
Alfvén, Hannes, 56, 61
America, 5, 116, 123, 181
Andersson, Lars, 133
Anshelm, Jonas, 50
antinuclear, 56–8, 60–1, 63–6, 69–73, 76, 153, 176
Arctic Circle, 28
Armenia, 86
Art Nouveau, 113
ASEA, 51
Ašmantas, Leonas, 91
atomic city. *See* mono-industrial town
atomograd. *See* mono-industrial town
authorized heritage discourse, 38
Avesta, 15–16, 103, 115, 118, **127–51**, 153–5, 157
 Jernverks AB, 128, **130–4**, 150
 mining and metal museum, 131–2, 150
 Polarit, 148
 sheet rolling mill, 133, 139–40, 142–4, 148
AXE, 55

Baltic Sea, 12–13, 62, 88, 95
Baltimore, Maryland, 16
Barsebäck, nuclear power plant, 13–14, **48–73**, 89–90, 92, 153–6
Barsebäck, spirit of, 64, 153
Barsebäck Seaside, 73, 155
Barsebäckshamn, 49
Barselina, 89–90
Bauman, Zygmunt, 9, 117
Berlin, 103–5, 120, 125
Berg, Kristian, 70
Bison, European, **132–4**, **148–9**
Boden, 24
Bosdotter, Kjersti, 145
Bottrop, 110
Brown, Kate, 93, 181–2
bruksanda. *See* spirit of the company town
Brussels, 75, 153
Burell, Jan, 146–7

Center Party, 56–7, 67, 144–5
Chernobyl, 41, 61–2, 66, 76, **82–3**, 88–9, 92, 95, 98, 156, 184
chocolate city, 78, 182
Cinis, Andis, 80
coal, 48, 63–4, 103–4, 113
Cold War, 94, 125
company land, 34–7, 42
company town, 15, 127, **133–5**, 138, 142, 148, 150–1, 153, 182
 see also mono-industrial town
Copenhagen, 14, 49, 59, 63–4, 68, 70–2, 156
Copper Valley, the, Avesta, 135
 see also Koppardalen
copper works, 15, 128, 135, 143
Cossons, Neil, 17

Dahl, Birgitta, 62–3
Dahlberg, Jonas, 4
Dalälven river, 128, 130
Danielsson, Tage, 61
Dawson, Jane I., 85
Degutis, Algimantas, 97
demonstration, 59–62, 65, 85
Denmark, 12–14, 48–9, 52–3, 55, **58–61**, 63–4, 68, 70–1, 73, 176
Detroit, Michigan, 19, 103, 114, 126, 155
dig-where-you-stand movement, 11
Drėmaitė, Marija, 80, 95
Drūkšiai, lake, 77, 84
Duisburg, 16, 102–3, 106, 110, 117–18, 120, 122, 153–4, 157, 191

E.ON, 67–9, 72–3
Ebert, Wolfgang, 106–7, 110
eco museum, 11
Eden, 93, 121, 155, 182
Edensor, Tim, 18, 118
Electricity Museum, Denmark, 71
Elektrenai, 81
Elkraft, Denmark, 52
Emscher river, 103, 111, 121
Energy and Technology Museum, Vilnius, 96
Estonia, 81–2, 86
Euratom, 89
European Bank for Reconstruction and Development (EBRD), 89–91
European Coal and Steel Community, 103
European Union (EU), 14, 75, 89–91, 94, 138, 154, 187
Everbrand, Lars Åke, 137–9, 141–3, 147

Fälldin, Torbjörn, 57
fear, 1, 12–14, 24, 47–8, 58–9, 61, 66–7, 73, 92, 94–5, 109, 156, 173
Föhl, Axel, 192

Foote, Kenneth E., 123
forest, forester, 41, 66, 77, 101, 119–21, 125, 128, 155
Forsberg, Gunnel, 150
Foss, Nicole, 90
France, 11, 51, 56, 59, 66, 195

Gällivare, 22–31, 33, 35–7, 43–5
  municipal museum, 35
Gandy, Matthew, 125
Ganser, Karl, 105, 123
garden, 18, 73, 109, 112, 116, 119–21, 125, 155
Gävle, 62
Gellivare Malmfält, AB, (AGM), 23–4
German Society for Industrial History, 106, 191
Germany, 12–13, 15, 25, 56, **101–26**, 188, 195
Giblin, John Daniel, 6
Gillhög, 52
Gorbachev, Mikhail, 86
Gothenburg, 17
Graham, Brian J., 5
Great Britain, 11, 51, 132, 195
Greenpeace, 65, 71
Greider, Göran, 19
Greifswald nuclear power plant, 62
Gustafsson, Fredrik, 30
Gylys, Jonas, 91

Hagerman, Chris, 115
Hague, La, 66
Hallman, Per Olof, 25
Harrisburg accident, 57, 61
  see also Three Mile Island incident
heal, healing, 2–3, 6, 11–12, 16, 18, 21, 73, 97, 124–6, 151, 154–5, 158
Henare, Amiria J. M., 8
Hermelin, Bo, 131–2, 150
history from below, Geschichte von unten, 11, 113–14, 132, 146–7
Höfer, Wolfram, 116
Holbraad, Martin, 8
Howard, Peter, 5

IAEA, 85
Ignalina nuclear power plant, 14–15, 62, 67, 75–99, 153–4, 185
industrial archeology, 11
Industriewald Ruhrgebiet, 121–2
Internationale Bauausstellung (IBA) Emscher Park, 15, 105–6, 110, 122–3
ironworks, iron and steel works, 7, 15–16, 102, 106–7, 110, 113, 115, 118, 120–1, 128–30, 134, 146, 149, 154, 191
Isacson, Maths, 146
Isling, Pär, 30

Järegård, Ernst-Hugo, 63
Johansson, Gunnar, 90
Johnson family (Avesta), 131–2, 134–6, 141, 146, 148
Jönsson, Lars-Eric, 70
Josephson, Paul, 87

Kaijser, Arne, 57
Kalm, Mart, 80
Karosta, 2
Kaunas, 81, 91
Kävlinge, 55, 63–4, 68–9, 73
Keil, Andreas, 111–12
Kiev, 62
Kift, Dagmar, 113
Kiruna, 13, 22, 24–5, 28, 30, 37–40, 45
  model town, 25, 37, 39
Koppardalen, 135–6, 138–42, 144–8, 150–1, 155
Kowarik, Ingo, 104, 119
Krohn Andersson, Fredrik, 52
Krupar, Shiloh R., 4, 114–15

Landschaftspark Duisburg-Nord, 15, 101–3, 106, 108–12, 117–18, 120, 124, 191
Latvia, 2, 76, 81–2, 86, 92, 98
Latz, Peter, 108–9
Latz und Partner, 108

Left Party, 56, 67
Lekan, Thomas, 116
Lithuania, 12–14, 67, 75–99, 156–7, 185–6
Lithuanian Academy of Sciences, 84
Lithuanian Movement for Perestroika. *See* Sąjūdis
LKAB, 24–5, 27–39, 42–5, 153
  company museum, 35
Löddeköpinge, 48–51, 54–5, 73
Löfwall, Ulf, 139, 143
Logan, William Stewart, 6
London, 17
Lowell, Massachusetts, 16
Lowenthal, David, 122
Luleå, 24–5
Lund, Anne, 59
Lundberg, Arne S., 27–9
Lundbohm, Hjalmar, 25, 37
LWL-Industriemuseum (Dortmund), 113

Macdonald, Sharon, 6
Malmberget, 12–13, 22–45, 73, 95, 153–4, 156–7
Meidericher Hütte, 106–8, 112
Mellander, Barbro, 70
Millington, Nate, 114–15
mining, 13, 21–35, 37–40, 42, 44–5, 103–4, 118, 131–2, 145, 150, 153–4, 167, 183
mining town, 12, 21–2, 28, 40, 45, 154
  see also company town, mono-industrial town
Minsredmash, 77, 81, 86, 88
model town (Kiruna), 25, 37, 39
mono-industrial town, 1, 69, 77–9, 95, 182
  see also company town
Moscow, 14, 75, 80, 83, 85, 153

Narvik, 24–5
National Heritage Board, 37, 52, 71
national interest, areas of, sites of, 34–5, 37–8, 132, 138

National Museum of Denmark, 71–2
National Museum of Science and Technology, Stockholm, 70, 132, 180
nature
 of a fourth kind, 104, 116, 119, 125
 industrial, 15, 101, 103–6, 108, 111–12, 114, 116–26, 155, 157–8, 189
 urban-industrial, 104–5, 116, 120, 125
New York, 16, 68
Nisser, Marie, 131
Nistad, Jan, 90
Norberg, Dan Ola, 141–3
Nordlund, Therese, 134
North Rhine-Westphalia, 102, 105, 113
Northern Works, Avesta, 129–31, 134–5, 138, 140–2, 147–8
nuclear waste. *See* radioactive waste
Nylund, Thomas, 38–9
Nyström, Tommy, 33, 41, 43

OKQ8, 144
Organisationen til Oplysning om Atomkraft, (OOA), **58–61**, 64, 68, 71
Öst, Leif, 65, 90
Outokumpu Steel, 148
overgrowing, 1, 15, 101, 109, 114–15, 119–20, 122, 126, 155, 157

palimpsest, 3–4, 125
Palme, Olof, 57
Palmqvist, Roland, 55
paradise, 14, 75, 78, 94, 99, 112, 121–2, 155, 182
Perers, Karin, 137, 144–5
perestroika, 83, 85, 94
Pit, the, Malmberget, 12–13, 22, 37, **40–2**, **44–5**, 154, 156–7
plutonium, 63, **80–1**, 93, 178

post-industrial society, 9
Pripyat, 82–3
probabilistic safety analysis, (PSA), 89

radioactive waste, 14–15, 47, 51, 57, 59, 75, 92, 94, 97
radioactivity, 47, 82, 92–3, 155
RBMK, 76, **80–2**, 87–9, 92, 98
RDS, 55
Reel Energi Oplysning, (REO), 59
Reeves, Keir, 6
referendum, about nuclear power in Sweden, 14, **57–8**, 67
reuse, 6–7, 15–19, 115, 146–7, 154–5, 188
Rhine river, 103
Ricœur, Paul, 99
Riget, TV series, 63
Ringhals nuclear power plant, 67
Risø research station, 50, 53, 59, 71
ruderals, ruderal species, 108, 158, 192
Ruhr, 15–16, 25, **101–26**, 155, 157–8, 192, 194
river, 103
ruin, ruined, ruination, 7, 16, 18–19, 98, 108, 118, 136, 155
Russia, Russian, 7, 14, 24, 75, 79, 81–2, 85, 87, 89, 91, 94, 97, 183, 186

Sąjūdis, Lithuanian Movement for Perestroika, 83, 85, 88
Sami, 23, 35, 38, 45
scab, 2, 5, 154
scarification, 21
Schöneberger Südgelände, Berlin, 120
Ševaldin, Viktor, 96–7
sheet rolling mill, Avesta, 133, 139–40, 142–4, 148
Sigyn, 71
Sjöholm, Jennie, 38–9
Skansen open air museum, Stockholm, 132
slag stone, 130, 137, 139–40, 148, 151
Šliavaitė, Kristina, 79, 86

INDEX

Sloss Furnaces, Birmingham, United States, 117
Smith, Laurajane, 8, 11, 38
Sniečkus, 14, 75-80, 85-7, 92, 94, 98, 153, 156, 186
Social Democratic Party, 57, 61, 63, 67, 105, 145
Sosnovjy Bor nuclear power plant, 62
Sound, the, (Öresund), 14, 48-50, 53, 58-9, 61, 68, 70, 73, 156
Southern Works, Avesta, 129-30, 133-4, 148
Soviet Union, Soviet, 2, 14, 50, 62-3, **75-89**, **91-6**, **98-9**, 156-7, 181-4
Soyez, Dietrich, 6
spirit of Barsebäck, 64, 153
spirit of the company town, 133, 142, 148, 150-1, 153
Steen Nielsen, Jørgen, 61
sun badge, **59-60**, 71-2
Svenskt Stål AB, (SSAB), 134
Swedish Path, 50-1
Sydkraft, 48-56, 58, 62-4, 66-7, 171

Tate Modern, London, 17
Thamsten, Jan, 144
Three Mile Island incident, 57, 61, 88
Thyssen AG, 106
trade union, 61, 63, 95
Trier, Lars von, 63

Ukraine, 82, 86, 98, 156
undefined, post-industrial landscape scars, 7, 16, 19, 154
UNESCO, 18, 113
see also World Heritage List
United States, US, 11, 16, 19, 50-1, 56-7, 68, 88, 93-4, 103, 114-17, 123, 182-3

Urbonavičius, Saulius, 97-8
Urry, John, 9-10, 110
utopia, utopian, 12, 14-15, 36, 44, 75, 77, 83, 99, 182
Utøya massacre, 4-5

Vaišvila, Zigmas, 83
VATESI, 88
Vikhög, 49, 54
Vilnius, 75, 96
Visaginas, 75-6, 87, 91-4, 98, 153, 156
VNIPIET, All-Union Scientific Research and Design Institute for Energy Technologies, 80
Völklingen ironworks, 113, 121
VVER, 81, 87-8

Wastell, Sari, 8
welfare, 12, 26, 44-5, 71, 155-6
Welles, Orson, 56
Wetterlundh, Sune, 50
Wickman, Gustaf, 25
Woods, Lebbeus, 5
World Heritage List, 18, 101, 113, 118
World War I, 122
World War II, 16, 25, 35, 50, 80, 103-4, 120, 122, 129, 130, 183
wound, 1-5, 12, 16, 21, 27, 97-8, 110, 123-4, 136, 151, 154-6

Zeche Zollern II/IV, 113
Zeche Zollverein, 113
Zeller, Thomas, 116
Žemyna, environmental club, **83-5**, 88, 186
Zöpel, Christoph, 105
Zukin, Sharon, 17, 123